Synthesis Lectures on Ocean Systems Engineering

Series Editor

Nikolas Xiros, University of New Orleans, New Orleans, USA

The series publishes short books on state-of-the-art research and applications in related and interdependent areas of design, construction, maintenance and operation of marine vessels and structures as well as ocean and oceanic engineering.

Alexander Arnfinn Olsen

Cathodic Protection
of Marine Vessels

Alexander Arnfinn Olsen
Southampton, UK

ISSN 2692-4420   ISSN 2692-4471 (electronic)
Synthesis Lectures on Ocean Systems Engineering
ISBN 978-3-031-77577-2   ISBN 978-3-031-77578-9 (eBook)
https://doi.org/10.1007/978-3-031-77578-9

This Springer imprint is published by the registered company Springer Nature Switzerland AG
The registered company address is: Gewerbestrasse 11, 6330 Cham, Switzerland

If disposing of this product, please recycle the paper.

Preface

Protective coatings are the most efficient way to protect ship steel structures from corrosion. However, cathodic protection, often in conjunction with protective coatings, is also used to protect immersed parts of bare steel surfaces (including coating damaged areas) from corrosion. This includes the external hull surface and the internal surfaces of tanks, such as ballast tanks. Cathodic protection (CP) can be Impressed Current Cathodic Protection, Galvanic Anode Cathodic Protection, or a combination of both. Cathodic protection controls corrosion by supplying direct current to the immersed surface of the structure, thus making the structure a cathode of a cell. The external hull of a ship is exposed to different waters with differing chemistries, which have a profound influence on the cathodic protection.

This short text on *Cathodic Protection of Marine Vessels* was developed to provide guidelines for ship cathodic protection design, installation, and maintenance given it is a common practice for a ship to have cathodic protection systems installed during its new construction.

Southampton, UK   Alexander Arnfinn Olsen

Acknowledgements My deepest gratitude and thanks go to the amazing team at BAE Systems and Babcock Marine for their guidance, patience, and support. I also extend my thanks to the editorial team at Springer, and of course to my wife, without whose support and guiding hand this work would not be possible.

This title is published with the kind permission of the American Bureau of Shipping.

Contents

List of Figures

List of Tables

General

1

1.1 Scope

This short text has been written to define the best practices and design recommendations for a ship's corrosion protection using cathodic protection (CP) systems and provide guidelines for ship cathodic protection design, installation, and maintenance.

Protective coatings are the most efficient way to protect ship steel structures from corrosion. However, cathodic protection, often in conjunction with protective coatings, is also used to protect immersed parts of bare steel surfaces (including coating damaged areas) from corrosion. This includes the external hull surface and the internal surfaces of tanks, such as ballast tanks. It is a common practice for a ship to have cathodic protection systems installed during its new construction. External underwater surfaces include the underwater hull and all attachments and openings such as shafts, rudder, stabilisers, propeller, propeller bracket, sea chests, water intakes (up to the first valve), and scoops thruster). The internal surfaces of various tanks include those of ballast water tanks, freshwater tanks, fuel storage tanks, crude oil tanks, bilges, etc. Pipes, such as condensers and heat exchangers, should also be protected cathodically.

Cathodic protection (CP) can be Impressed Current Cathodic Protection, Galvanic Anode Cathodic Protection, or a combination of both. Cathodic protection controls corrosion by supplying direct current to the immersed surface of the structure, thus making the structure a cathode of a cell. The external hull of a ship is exposed to different waters with differing chemistries, which have a profound influence on the current distribution from the cathodic protection system. Thus, this text covers ships' cathodic protection only.

A. A. Olsen, *Cathodic Protection of Marine Vessels*, Synthesis Lectures on Ocean Systems Engineering, https://doi.org/10.1007/978-3-031-77578-9_1

1.2   Materials

This text covers hulls fabricated principally from carbon manganese steels or low-alloy steels and appurtenances made of other ferrous or non-ferrous alloys such as stainless steels and copper alloys. It considers issues from dissimilar metals on ships. The cathodic protection system should be designed so that there is complete control over any galvanic coupling. This text does not cover hulls principally made of other materials such as aluminium alloys, stainless steels, or concrete. The cathodic protection system for a ship is generally combined with a protective coating system, even though some appurtenances such as propellers are generally not coated. This text can be applied to both coated and bare hulls.

1.3   Personnel

Personnel responsible for the design, testing, measurements, monitoring, supervision of installation, supervision of operation, supervision of maintenance, and survey of cathodic protection systems are to have requite experience and qualifications in cathodic protection. Personnel are to be trained and certified to reach and demonstrate the necessary competence levels for the tasks undertaken. EN 15257 and the NACE Training and Certification Programme provide a suitable method of assessing and certifying the competence of cathodic protection personnel.

1.4   Normative References

The following documents, in whole or in part, are normatively referenced in this text:

EN 12473: General principles of cathodic protection in seawater.
EN 12496: Galvanic anodes for cathodic protection in seawater and saline mud.
EN 13509: Cathodic protection measurement techniques.
EN 15257: Cathodic protection personnel qualification.
EN 50162: Protection against corrosion by stray current from direct current systems.
EN 16222: Cathodic protection of ship hulls.
NACE SP0176: Corrosion control of submerged areas of permanently installed steel offshore structures associated with petroleum production.
NACE SP0492: Metallurgical and inspection requirements for offshore pipeline bracelet anodes.
NACE SP0387: Metallurgical and inspection requirements for cast galvanic anodes for offshore applications.

1.5   Terms and Definitions

For the purposes of this text, the terms and definitions are given in the following:

Anaerobic. Lacking free oxygen.

Anode (in general). An electrode, placed in the electrolyte, to apply cathodic protection to the structure.

Anode Potential. Anode-to-electrolyte potential.

Cathode. An electrode through which direct current leaves an electrolyte causing reduction reactions.

Cathodic Disbonding. Failure of adhesion between a coating and a metallic surface due to the effect of cathodic polarisation.

Closed Circuit Potential. Potential measured at a galvanic anode when a current is flowing in between the anode and the surface being protected.

Dielectric Shield. Alkali-resistant organic coating applied to the structure being protected in the immediate vicinity of an impressed current anode to enhance the spread of cathodic protection and minimise the risk of hydrogen damage to the protected structure in the vicinity of the anode.

Driving Voltage. Also called driving potential. It is the difference between the structure/electrolyte potential and the anode/electrolyte potential when the cathodic protection is operating.

Electrolyte. A liquid in which electric current may flow by the movement of ions.

Impressed Current (IC). Direct current supplied by an external power source to cathodically protect a structure.

Impressed Current Anode. Anode in an impressed current installation, which is connected to the positive terminal of an impressed current power supply.

Polarisation. Change in the potential of an electrode as the result of current flow to or from that electrode. Also, a change in the potential of a corroding metal from its natural steady-state value, as a result of current flow. "Depolarisation" is the reverse process of Polarisation.

Reference Electrode. An electrode which has a stable potential in one or more electrolytes at a given temperature, thus enabling it to be used in the measurement of other electrode potentials.

Silver/Silver Chloride (Ag/AgCl) Reference Electrode. Reference electrode consisting of silver, coated with silver chloride, in an electrolyte containing chloride ions. These are used mainly with seawater.

Stray Current. Current flowing through paths other than the intended circuit.

Underwater Hull. Part of the underwater hull may include that below the light waterline.

Design Criteria and Recommendations

2

2.1 General

A cathodic protection system uses galvanic anodes, an impressed current system, or a combination of both. The cathodic protection system should provide sufficient and well-distributed currents to the ship hull steel surfaces so that the surfaces can be polarised to the potential within the limits given by the protection criteria (refer to this chapter, Sect. 2.3) over the design life (refer to this chapter, Sect. 2.2). The potential should be as uniform as possible over the entire submerged surface. This objective may only be approached by adequate distribution of the protective current over the structure during its normal service conditions. Special considerations should be given for areas such as water intakes, thrusters, and sea chests.

The design of the cathodic protection systems for the external hull should consider the anticipated service conditions and operating conditions such as the characteristics of the seawater (e.g., brackish/freshwater, temperature), the average and maximum anticipated speeds, and the lifetime associated with static (berthed) and dynamic (sailing) conditions. The design of the cathodic protection systems for internal surfaces of ballast tanks should also consider the ballast period (wet/dry alternating period) and the influence of the ballast water treatment system.

2.2 Design Life of Cathodic Protection Systems

The cathodic protection system should be selected and designed either for the design life of the ship or on the basis of expected dry docking intervals, which are normally specified by the owner. The design life should take into account any period of time when the cathodic protection system is active, including the ship's construction period.

© The Author(s), under exclusive license to Springer Nature Switzerland AG 2025
A. A. Olsen, *Cathodic Protection of Marine Vessels*, Synthesis Lectures on Ocean
Systems Engineering, https://doi.org/10.1007/978-3-031-77578-9_2

2.3   Cathodic Protection Potential

2.3.1   Cathodic Protection Potential Criteria

The accepted criterion for protection of carbon steels or low-alloy steels in aerated seawater is a protection potential of −0.80 V or more negative measured with respect to Ag/AgCl/seawater reference electrode. With such polarisation, the corrosion rate of the steel in seawater is suppressed to an acceptably low level. These criteria have been developed through laboratory experiments and field experience. However, structure protection does not need to be limited to these criteria if it can be demonstrated by other means that control of corrosion has been achieved. In the case of mild steel with active sulphate-reducing bacteria (generally in anaerobic conditions), the potential for protection is −0.90 V instead of −0.80 V with respect to Ag/AgCl/seawater reference electrode. With increasing negative potentials, there may be an adverse effect on fatigue properties and a risk of hydrogen embrittlement of susceptible steels. Refer to this chapter, Sect. 2.3.2. A summary of the recommended potentials for protection of the various materials in seawater is given in Table 2.1.

The above potential criteria and limit values are "polarised" and do not include the voltage (IR) drop, which is generally considered insignificant in marine applications.

2.3.2   Detrimental Effects from Cathodic Protection

2.3.2.1   Protective Coatings
The coatings to be used should be prequalified for resistance to cathodic disbonding.

2.3.2.2   Mild Steels
With increasing negative potentials, there is an adverse effect on fatigue properties and a risk of hydrogen embrittlement of susceptible steels. For mild steel, a negative limit of −1.10 V (Ag/AgCl/seawater reference electrode) is generally accepted as the negative limit.

2.3.2.3   High-Strength Steels
For high-strength steels with yield strengths > 690 MPa (100 ksi) or hardness > 350 HV, it has been the practice to use potentials in the range of −0.80 to −0.95 V (Ag/AgCl/seawater reference electrode). For high-strength steels susceptible to hydrogen-induced stress cracking (HISC), the maximum negative potential should be more positive (less negative) than −0.83 V (Ag/AgCl/seawater reference electrode).

Table 2.1 Summary of potentials versus Ag/AgCl/Seawater reference electrode recommended for the cathodic protection of various metals in seawater

Material	Minimum negative potential volts	Maximum negative potential volts
Iron and steel	−0.80 for aerobic environment	−1.10
	−0.90 for anaerobic environment (with active sulphate reducing bacteria)	−1.10
High-strength steels (yield strength > 690 MPa or hardness > 350 HV)	−0.80	−0.83 to −0.95 [1]
Austenitic stainless steel for aerobic and anaerobic conditions N_{PRE} = % Cr + 3.3% (Mo + 0.5W) + 16% N	−0.30 for $N_{PRE} \geq$ 40 [2]	−1.10
	−0.60 NPRE < 40 [2]	−1.10
Duplex stainless steel for aerobic and anaerobic conditions	−0.60 [2]	Refer to Note 3
Martensitic stainless steel (13% Cr) for aerobic and anaerobic conditions	−0.50	Refer to Note 5
Nickel-based alloys	−0.20	Refer to Note 4
Copper alloys	−0.45 to −0.60 for alloys with aluminium	−1.10
	−0.45 to −0.60 for alloys without aluminium	No limit

Note (1) For high-strength steels susceptible to hydrogen-induced stress cracking (HISC), the maximum negative potential should be more positive (less negative) than −0.83 V (Ag/AgCl/seawater reference electrode). (2) For most applications, these potentials are adequate for the protection of crevices, although higher (more negative) potentials may be considered. (3) Forgings, castings, and welds are more prone to HISC than wrought materials due to the coarse microstructure allowing HISC propagating preferentially in the ferritic phase. (4) High-strength nickel copper and nickel chromium iron alloys can be subject to HISC, and potentials that result in significant hydrogen evolution should be avoided. (5) Depending on strength, specific metallurgic condition, and stress level encountered in service, those alloys can be susceptible to hydrogen embrittlement and cracking. If a risk of hydrogen embrittlement exists, then potential more negative than −0.80 V should be avoided

2.3.2.4 Austenitic Stainless Steels and Nickel-Based Alloys

Austenitic stainless steels and nickel-based alloys in the solution annealed condition are generally considered immune to HISC for all practical applications. Moderate cold work would not induce HISC sensitivity of these materials except for UNS S30200 (AISI 302) and UNS S30400 (AISI 304) stainless steels. The same applies for welding or hot forming according to an appropriate procedure. For certain nickel-based alloys, precipitation hardening may induce increased sensitivity to HISC. For precipitation hardened austenitic stainless steels, the susceptibility is low and a hardness of maximum 300 HV may be considered a reasonably safe limit, while component materials with a hardness higher than 350 HV should generally not receive cathodic protection. In the intermediate hardness range (i.e., 300–350 HV), precautions should be taken during design to avoid local yielding and/or to specify a qualified coating system for eliminating hydrogen absorption by cathodic protection. The qualified coating system should resist cathodic protection disbanding during service.

2.3.2.5 Ferritic and Ferritic-Pearlitic Steels (Stainless Steels)

Ferritic and ferritic-pearlitic structural steels with specified minimum yield strength (SMYS) up to 500 MPa (72 ksi) have proven in practice to have compatibility with marine cathodic protection systems. However, laboratory testing has demonstrated that steels which have passed their yield point have susceptibility to HISC. It is recommended that all welds should have 350 HV of hardness as an absolute upper limit. Within the range of 300–350 HV hardness, precautions should be applied during design to avoid local yielding and/or to specify a qualified coating system for eliminating hydrogen absorption by cathodic protection. Again, the qualified coating system should resist cathodic protection disbonding.

2.3.2.6 Martensitic Carbon, Low-Alloy, and Stainless Steels

For martensitic carbon, low-alloy, and stainless steels, failures of cathodic protection-induced HISC have occurred in materials with a yield stress of 690 MPa (100 ksi) and a hardness of 350 HV. It is widely recognised that untempered martensite is prone to HISC. Welds susceptible to martensite formation should receive post-welding heat treatment (PWHT) so as to reduce heat affected zone (HAZ) hardness and residual stresses from welding. The same recommendations for hardness limits and design consideration for ferritic steels above apply.

2.3.2.7 Ferritic-Austenitic ("Duplex") Stainless Steels

Ferritic-austenitic ("duplex") stainless steels should be regarded as potentially susceptible to HISC, independent of SMYS [typically 400–550 MPa (58–80 ksi)] or specified maximum hardness. Welding may cause increased HISC susceptibility in the weld metal and in the HAZ adjacent to the fusion line. This is related to an increased ferrite content rather than hardness. Qualification of welding should therefore prove that the maximum ferrite content in the weld metal and the inner HAZ (about 0.1 mm (0.004 in) wide) can be efficiently controlled; contents of maximum 60–70% are typically specified. Forgings are more prone to HISC than wrought materials due to the coarse microstructure allowing HISC to propagate preferentially in the ferrite phase. Cold bent pipes with small diameters [uncoated and with mechanical connections (i.e., no welding)] have a proven record for cathodic protection compatibility when used as production control piping for subsea installations. Design precautions should include avoiding local plastic yielding and use of qualified coating systems.

2.3.2.8 Copper- and Aluminium-Based Alloys

Copper- and aluminium-based alloys are generally considered immune to HISC, regardless of fabrication modes.

2.3.3 Potential Measurements

The protection criteria and effectiveness of cathodic protection systems should be confirmed by direct measurement of the structure potential. However, visual observations of progressive coating deterioration and/or corrosion are indicators of possible inadequate protection. Steel plate thickness gauging can also indicate deficiencies in corrosion protection. Potential measurements should be made with the reference electrode and a high impedance (minimum 10 MΩ). The reference electrode should be submerged within the seawater as close as practicable to the ship's hull to minimise voltage drops. The following should be considered when measuring the hull potentials:

(1) Potential measurement techniques can be referred to EN 13509 "Cathodic Protection Measurement Techniques"
(2) In areas of greatest shielding and/or in brackish waters, particular attention should be given in evaluating the protective level of ship hulls
(3) Changes in water resistivity from causes such as freshwater flow from a river or temperature variation can affect voltage level
(4) In impressed current systems under conditions involving high-resistivity water or high current density, the voltage drop may be excessive. Instant-off potential measurements can provide useful information by eliminating voltage drop in the water by switching off the direct current (DC); and
(5) In conventionally designed galvanic anode protection systems, current-off readings are not possible. However, the included voltage drop is generally not significant in ordinary seawater if the reference electrode is placed close to the structure. The voltage drop may become significant in brackish waters. In such cases, it may be necessary to use interruptible coupons or other IR correction techniques (refer to NACE SP0169) to determine the true potential of the metal surface.

When permanent measurement electrodes are installed, they should be at locations representative of the most negative and the most positive potentials on the external hull surface. While the exact electrode locations are known, the information obtained from these electrodes is limited to the adjacent structure surfaces. Although this limitation holds true for any potential measurement, this method can provide a reproducible basis for comparing potentials at different times. The accuracy of permanent electrodes should be periodically checked against another electrode. Dual reference electrodes that combine zinc and Ag/AgCl/seawater references into a single, permanently installed unit also help to detect/reduce malfunctions.

In the case of impressed current systems, reference electrodes should be fitted to the structure at suitable locations in order to automatically control the output of the anodes and maintain the polarisation of critical areas within the set limits. If this measurement circuit remains permanently connected, care should be taken that it does not draw excessive current (including when idle) from the reference electrode, which may become polarised and give false indications.

The use of portable coupons and electrical resistance probes can be considered for commissioning to allow readings to be made in areas where there is no permanent reference electrode installation. They can be particularly useful when establishing the effectiveness of the dielectric shields. In addition to reference electrodes, some structures are equipped with permanent monitors to measure current density and current output from representative galvanic anodes. These devices are particularly useful when assessing new structure designs or new environments in which precise cathodic protection design criteria are not available. These devices typically use calibrated shunts to arrive at the current output or current density value. Signals are usually transmitted topside using hard-wired connections.

2.3.4 Reference Electrode

The polarised potential of the steel surface in its environment (such as seawater) is measured relative to the potential of a reference electrode. Consequently, no equipment item is of more fundamental importance to the cathodic protection (CP) system than the reference electrode. The most commonly used reference electrode in marine environments is the solid junction (SJ) silver/silver chloride (SSC) electrode in direct contact with seawater. However, the potentials of a SJ Ag/AgCl/seawater electrode can be affected by the chloride concentration and temperature of the seawater in which it is immersed. It is also affected by light (photo-sensitive) and contaminants in its environments. These are most likely to affect reference electrodes which operate in a variety of environments and permanent electrodes on ships sailing in different waters (seawater, brackish, and fresh water). The reference potential will vary as the logarithm of the concentration of active species, which is especially significant in dry electrodes. This effect is minimised by way of wet electrodes. Temperature has direct and indirect effects on the reference electrode. The direct effect is a linear variation of potential with temperature [provided with temperature coefficient—mV/°C (mV/°F)]. The indirect effect is a function of the increased quantity of salt able to be dissolved at higher temperatures. These temperature effects are large enough to produce a significant error in potential measurements if left uncompensated. The effect of temperature on reference electrode potential can be calculated by the following equation:

$$E_t = E_{SHEat25\ °C} + k_t(T - 25\ °C)$$

$$E_t = E_{SHEat77\ °F} + k_t(T - 77\ °F)$$

where

E_t = reference potential at temperature, T
$E_{SHEat25°C}$ = reference potential at 25 °C (refer to Table 2.2)
$E_{SHEat77°F}$ = reference potential at 77°F (refer to Table 2.2)
k_t = temperature coefficient, in mV/°C (mV/°F)
T = actual temperature, in °C (°F).

Table 2.2 Common reference electrodes and their potentials and temperature coefficients

Electrode	Potential relative to SHE [at 25 °C (77 °F)] mV	Protection potential reading [at 25 °C (77 °F)] mV	Temperature coefficient kt mV/°C (mV/°F)	Typical usage
CSE (Cu/ CuSO$_4$/ Saturated CuSO$_4$)	+316	−860	+0.90 (+0.5)	Soil, fresh water
SSC (SJ Ag/ AgCl/0.6 M (3.5%) NaCl)	+256	−800	−0.33 (−0.18)	Seawater, brackish
SSC (LJ Ag/ AgCl/0.5 M KCl)	+256	−800	−	−
SSC (LJ Ag/ AgCl/KCl saturated)	+199	−743	−0.70 (−0.39)	−
SCE (Saturated Calomel Electrode)	+0.244	−788	−0.70 (−0.39)	Water, laboratory
Steel protection potential	−544	−	−	−
ZRE (Zinc Reference Electrode)	−780	+ 244	−	Seawater

Other standard reference electrodes may be substituted for the Ag/AgCl (seawater) with their protection potential equivalent to −800 mV referred to an Ag/AgCl (seawater):

- Saturated copper/copper sulphate reference electrode (CSE [saturated CuSO₄]): −860 V (or more negative) for protection. This electrode is not stable for long-term immersion service; and
- High-purity zinc electrode.

2.3.5   Factors Affecting Cathodic Protection Potential

Ships are exposed to water under varying temperatures and environments. The potential at which corrosion is controlled is a function of the temperatures and the environments. In seawater exposed to the air and at typical ambient temperatures, the criterion listed in this chapter, Sect. 2.3.1 has proved to be satisfactory. For other circumstances, the potential to control corrosion can be estimated using the Nernst equation:

$$ E = E^\circ + \frac{RT}{nF} \ln\left(\frac{a_M^{+ne}}{a_m}\right) $$

where

E° = standard state potential
E° = electrode potential
a_M^{+ne} = activity of metal ions
a_m = activity of metal (1 for pure metal)
R = gas constant (8.31431 J/($^\circ$K • mole))
T = absolute temperature (298.2°K)
n = number of electrons transferred
F = Faraday's constant (96,500 Coulombs/equivalent).

Factors that affect the degree of cathodic protection needed:

- *Water Velocity.* Cathodic protection current needs increase with water velocity past the hull. The necessary current can be as high as 30 times that for still water
- *Vessel Usage.* More frequently operated vessels need more cathodic protection than vessels infrequently used
- *Conductivity of the Water.* As the conductivity increases, the rate of galvanic activity increases
- *Salinity (Fresh and Seawater).* Current needs increase with salinity, but higher driving potentials are needed in fresh water

- *pH of the Water.* As pH decreases (acid rain lakes), the corrosion rate increases
- *Dissolved oxygen content*
- *Temperature*
- *Marine growth*; and
- *Deterioration of Protective Coatings.* Current needs increase as protective coatings deteriorate.

The above factors are interrelated and vary with geographical location and season. It is not feasible to give an exact relation between the seawater environmental parameters indicated above and cathodic current demands to achieve and to maintain cathodic protection. To rationalise cathodic protection design for marine applications, default design current densities, i_c, are defined based on vessel speed.

2.4 Design Current

2.4.1 General

To achieve the protection potential criteria denoted in this chapter, Sect. 2.3, the appropriate design current density for each component or area with respect to the environmental and service conditions should be used during design. The selection of design current densities may be based on either experience gained from similar ships operating in a similar manner or specific tests and measurements. The current demand of each metallic component or area of the structure is the result of the product of its surface area multiplied by the protection current density for the bare steel and the coating breakdown factor.

2.4.2 Structure Subdivision and Surface Area Calculations

2.4.2.1 Submerged Surfaces of Ship Hull

In the cathodic protection design, the submerged surfaces of a ship hull can be divided into different cathodic protection zones, which are then considered to be independent for cathodic protection design, although they are not electrically isolated. For instance, the underwater hull can be divided into two main cathodic protection zones: the forward (or bow) zone and the aft (or stern) zone, shown in Fig. 2.1. This subdivision is related to the high current demand of the aft zone due to high water flow rates, turbulence, and the presence of dissimilar metals due to the propeller(s). The aft cathodic protection zone includes the aft part of the hull, propeller(s), shaft(s), rudder(s), etc. Some specific components may constitute a cathodic protection zone, such as openings of sea chests, thrusters, rudders, and propellers. The cathodic protection system is determined and dedicated for each cathodic protection zone. Each component should be fully detailed in the design

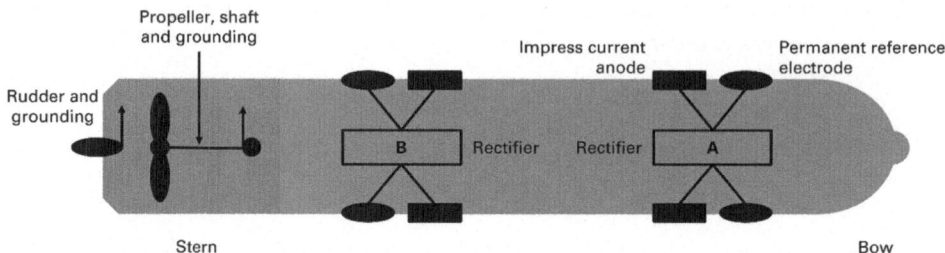

Fig. 2.1 Typical arrangement for two impressed current systems for ship hulls

including material, surface area and coating characteristics (coating type, design life and coating breakdown factor). Special considerations should be given for areas such as water intakes, thrusters, and sea chests:

- Electrochemical marine growth prevention systems (MGPS) often used within sea chests to prevent fouling of seawater intake systems. They may interact with the cathodic protection system. This effect should be considered in the design and installation of the anti-fouling system; and
- Cathodic protection within sea chests may adversely affect box coolers in sea chests if the box coolers are electrically isolated from the sea chest. The possible effect should be taken into account in designing the cathodic protection for sea chests.

2.4.2.2 Internal Surfaces

For internal surfaces, complex geometries can exist within tanks (e.g., stiffeners and heating coils). Lower sections of tanks not fully drained within stiffeners can also constitute discrete cathodic protection zones to be considered. Each component of a cathodic protection zone should be fully detailed in the design. This should include:

- Material type
- Specific potential limit (if applicable)
- Complexity of the structure
- Surface area; and
- Coating characteristics, including type, predicted lifetime, anticipated coating breakdown. Having component areas in the zone and current density needed regarding the component coating condition and service condition, the current demand for each zone can be determined. It is acceptable to simplify calculations of the surface areas of complex geometries, provided that the simplification is conservative. All calculations of surface areas and current demand should be documented in the cathodic protection design report.

Table 2.3 Typical design current densities

Situation m/s (knots)	Design current densities for bare steel mA/m^2 (mA/ft^2)	Design current densities for coated steel mA/m^2 (mA/ft^2)
V ≤ 1 (2 knots)	100–200 (9.3–18.6) without tidal influence	5–15 (0.5–1.4)
	150–250 (13.9–23.2) with tidal influence	7–20 (0.7–1.9)
1 < V < 10 (20 knots)	220–350 (20.4–32.5)	11–28 (1.0–2.6)
V ≥ 10 (20 knots)	350–500 (32.5–46.5)	18–40 (1.7–3.7)
Vessels in ice	500–750 (46.5–69.7)	35–90 (3.3–8.4)
Propeller surface	≥ 500 (46.5)	

2.4.3 Design Current Density for Bare Steel

The protection current density of bare steels and other metals (i.e., cuprous alloys) depends on the kinetics of electrochemical reactions on the surface. It varies with the material, potential, surface condition, dissolved oxygen content in seawater, flow rate, sailing speed, and temperature. For each environmental and service condition, protection current density should be evaluated. As guidance, typical design current densities for protecting bare steel are given in Table 2.3.

Protective current density for the areas of bow thrusters, water-jet drives, etc., should be determined individually in each case. In addition to the calculated underwater hull area, other cathodic protection zones (such as the appendages, propellers, shafts, sea cheats, thrusters, etc.) should be calculated separately.

2.4.4 Design Current Density for Coated Steel

The cathodic protection system is generally combined with suitable protective coating systems, which act as the primary method for corrosion prevention. Most protective coatings are organic and not conductive. These coatings can reduce the protection current demand tremendously and can also improve the current distribution over the surface. However, the coatings deteriorate with time because of aging and physical damage, and thus the protection current demand of coated steel also increases with time. An initial coating breakdown factor is used to present initial coating damages and defects when the ship is put into service. An annual coating deterioration rate is used to describe an increase of the coating breakdown factor over time during the design life of the ship's cathodic protection system or during a period corresponding to the drydocking interval (at least 2 years). For each individual component or cathodic protection zone (area and condition), the design current density needed for the protection of coated steel, J_c, is equal to the product of the current density for bare steel (refer to Chapter 2, Sect. 3.2), J_b, and the coating breakdown factor, f_c:

$$J_c = J_b \times f_c \quad \text{A/m}^2 \left(\text{A/ft}^2\right)$$

where

J_c = protection current density for coated metal, in A/m^2 (A/ft^2)
J_b = protection current density for bare metal, in A/m^2 (A/ft^2)
f_c = coating breakdown factor which varies with time due to ageing and mechanical damage
f_c = 0 for a perfectly insulating coating and
f_c = 1 for no coating or a coating with no insulation properties.

This formula should be applied for each individual component or cathodic protection zone. The fc values are strongly dependent on the actual construction and operational conditions. For reference, guidelines for the values of coating breakdown factors for conventional paint systems are given in Table 2.4. For high-speed vessels (speed above 25 knots) and vessels in harsh or ice service, the annual depletion rate may be higher.

Another approach for the estimation of the protection current density for coated structures may be considered when values for design parameters are well known from experience. In this approach, an average value of this protection current density (J_g) is considered. Detailed documentation regarding the class of vessel and service and the satisfactory cathodic protection performance is needed to be evaluated for this approach. For reference, typical average current densities for cathodic protection of coated hulls of ships are given in Table 2.5.

Table 2.4 Values of coating breakdown factors for conventional paint systems for ship hulls

Initial coating breakdown factor		1–2%
Indicative annual deterioration rates for durability defined ISO 12944 for seawater or brackish water immersion service	Using low durability coating systems	3% per year
	Using medium durability coating systems	1.5% per year
	Using high durability coating systems	0.5–1% per year

Table 2.5 Typical average current densities of coated hulls for cathodic protection

Docking periods months	Current densities mA/m^2 (mA/ft^2)
Up to 18	15–25 (1.4–2.3)
19–36	26–45 (2.4–4.2)
37–60	46–75 (4.3–7.0)

2.4.5   Current Demand

The current demand for each zone can be determined once component areas and current densities are available. The design of impressed current systems is based on the maximum current demand. The design of galvanic anodes systems is based on the maximum current demand and mean current demand. If typical average current densities in Table 2.5 are not adopted for the design, two different values, maximum current demand $I_{\max}$ and mean current demand I_{mean}, should be considered:

- I_{mean} is used to calculate the minimum mass of galvanic anode material or life of impressed current anodes necessary to maintain cathodic protection throughout the design period; and
- $I_{\max}$ corresponds to the most severe working conditions (e.g., dynamic conditions, end of life coating breakdown factor, and worst-case environmental conditions) and is used to design the maximum current capacity of the cathodic protection system.

For each component, these two protection current demands can be determined according to the following formula:

$$I_{i,\max} = S_i \times f_{c,\max} \times J_{bd} \quad \text{amperes}$$

$$I_{i,mean} = S_i \times f_{c,mean}[t \times J_{bd} + (1-t) \times J_{bs} \quad \text{amperes}$$

where

$I_{i,max}$ = maximum protection current demand for a component, in amperes (A)
$I_{i,mean}$ = mean protection current demand for a component, in amperes (A)
S_i = area of the submerged zone (component under full load conditions including the underwater hull and boot topping), in m^2 (in^2)
$f_{c,max}$ = maximum coating breakdown factor for the concerned service period
$f_{c,max}$ = 1 for bare steel surface
$f_{c,mean}$ = mean coating breakdown factor for the concerned service period
J_{bd} = current density for bare metal in dynamic conditions, in A/m^2 (A/in^2)
J_{bs} = current density for bare metal in static conditions, in A/m^2 (A/in^2)
t = fraction of time associated to dynamic conditions.

Consequently, for each cathodic protection zone, the protection current demand values are given by the sum of all components in the zone:

$$I_{\max} = \sum I_{i,\max}$$

$$I_{mean} = \sum I_{i,mean}$$

When the average current density approach for each component is used, referring to Table 2.5, a unique current demand $I = J_g \sum S_i$ is considered for each cathodic protection zone.

2.5   Circuit Resistance

The number and location of the anodes determines an electrical current distribution for achieving the protection potential level over the whole steel structure surface. The potential drop from cathodic protection circuit resistance should be considered for applied potential to the steel structure surface. For cathodic protection systems using galvanic anodes, the optimum anode dimensions may be determined using Ohm's law of cathodic protection circuit:

$$I = \frac{\Delta E}{R}$$

where

I = current output from anode, in amperes
ΔE = driving voltage between anode and structure, in volts
R = circuit resistance, which is sum of anode-to-electrolyte resistance, electrolyte resistance and structure-to-electrolyte resistance, in ohms.

The circuit resistance, R, is assumed to be approximately equal to the anode/electrolyte resistance, which is called "anode resistance", as the structure-to-electrolyte resistance and the electrolyte resistance are generally very small. ΔE is generally taken as the potential difference between the polarised potential of the steel and the operating potential of the particular anode alloy in seawater.

The anode resistance is a function of the anode geometry and the resistivity of the electrolyte. Empirical formulae given in the following Subsection may be used to calculate the anode resistance. For an impressed current system, the DC output voltage of the power source is to be higher than the sum of the voltage drops in all the components of the circuit cables, electrolyte resistance, the anode-to-electrolyte resistance and structure-to-electrolyte resistance. The voltage between impressed current cathodic protection (ICCP) anode and electrolyte should not exceed the maximum acceptable value depending on the material of the anode.

2.6 Anode Resistance Calculations

For galvanic anode cathodic system, the anode/electrolyte resistance (called "anode resistance") approximately represents the circuit resistance, as the structure-to-electrolyte resistance and the electrolyte resistance are generally very small. The electrical resistance of an anode to the surrounding electrolyte depends upon electrolyte resistivity and on the size of shape of the anode. Empirical formulae given in this chapter, Sect. 2.6.1 through Sect. 2.6.3 may be used to calculate the anode resistance. For high resistivity electrolytes, anodes in form of rods, slender ingots, or strip/ribbon should be used in order to reduce anode resistance. For low resistivity electrolytes, thicker rods or ingots may be used for giving an adequate life. For closely spaced arrays of anodes:

(1) Anodes in close proximity will affect the electrical field around adjacent anodes and reduce the current output from anodes. Effectively, the resistance of the individual anode in an array anode will be increased by the proximity to adjacent anodes
(2) Closely spaced anodes should be the subject of specific design assessment and their resistance may be determined by using alternate classical resistance to earth formulae, mathematical modelling, or direct field measurement; and
(3) If anodes are grouped close to each other, mutual interference between anodes should be considered when calculating the anode resistance.

2.6.1 For Slender Anodes Mounted at Least 0.3 m (11.8 in) Offset from the Structure Steel Surface

If $L \geq 4r$:

$$R_a = \frac{\rho}{2\pi L}\left[1n\left(\frac{4L}{r}\right) - 1\right] \ \text{Ohms}$$

If $L \geq 4r$:

$$R_a = \frac{\rho}{2\pi L}\left\{1n\left[\frac{2L}{r}\left(1 + \sqrt{1 + \left(\frac{r}{2L}\right)^2}\right)\right] + \frac{r}{2L} - \sqrt{1 + \left(\frac{r}{2L}\right)^2}\right\} \ \text{ohms}$$

where

    L = length of the anode, in m (in)
    R = radius of the anode, in m (in)
    ρ = seawater resistivity, in ohm-m (ohm-in).

For anodes mounted < 0.3 m (11.8 in) and > 0.15 m (5.9 in) offset from the steel surface, the resistance can be assumed to be $R_a \times 1.3$.

2.6.2 Long Flush Mounted Anodes on the Structure Steel Surface Where Length ≥ 4 × Width

The following equation is typically applicable to long hull anodes (impressed current or galvanic):

$$R_a = \frac{\rho}{2S} \quad \text{Ohms}$$

where

    ρ = seawater resistivity, in ohm-m (ohm-in)
    S = arithmetic mean of anode length and width, in m (in).

2.6.3 Short Flat Plate Mounted Flush on the Structure Steel Surface Where Length < 4 × Width

The following equation is typically applicable to circular bow-mounted impressed current anodes or short galvanic anodes:

$$R_a = 0.315 \frac{\rho}{\sqrt{A}} \quad \text{Ohms}$$

where

    ρ = environment resistivity, in ohm-m (ohm-in)
    A = exposed surface area of anode, in m^2 (in^2).

2.7   Anode Current Output and Operating Life

2.7.1   General

The current output of an anode depends on seawater resistivity and anode dimensions. The specific consumption rates also highly depend on its operation environments. The anode life depends on the anodic material's consumption rate and its weight for a given electrical current output. The dimensions and number of anodes and the distribution of anodes should be optimised in order to minimise the total weight of the galvanic anodes and to provide a protective electrical current greater or equal to the mean and maximum protection current demands for the life of the anodes.

2.7.2   Service Temperature Effect

In the case of ambient temperatures exceeding 25 °C (75 °F), the reduced capacity and effectiveness of the sacrificial anodes should be taken into account for the design and arrangement. This is especially applicable to hot transverse bulkheads (e.g., walls adjoining fuel tanks). Conventional sacrificial anodes of zinc are only to be used up to an ambient temperature of 50 °C (122 °F) for the protection of steel. If special alloys are to be used at temperatures exceeding 50 °C (122 °F), their electrochemical characteristics and protective effect should be verified separately. The capacity of aluminium anodes is also reduced with temperature increase. In the case of temperatures $T = 20\text{--}80$ °C (68–176 °F), the current capacity can be calculated as an approximation from using the following equation:

$$Q(T) = 2000 - 27 \times T - 20 \,°\text{C A - h/kg}$$

$$Q(T) = 907 - 6.8 \times T - 68 \,°\text{F A - h/lb}$$

Experience shows that there are also special alloys for aluminium anodes which possess greater current capacities at high temperatures than the values calculated from the above equation. The manufacturer should verify and provide these values.

2.7.3   Minimum Net Weight of Anode

The minimum total net weight of the anode material needed for a cathodic protection zone may be determined from:

$$W_{total} = \frac{(I_{mean}) \times T_{design} \times 8760)}{Q \times u} \,\text{kg}$$

$$W_{total} = \frac{2.2046 \times (I_{mean}) \times T_{design} \times 8760)}{Q \times u} \text{lbs}$$

where

W_{total} = minimum total net weight of galvanic anode material needed, in kilograms (kg) [pounds (lb)]

$(I_{mean}$ = total maintenance current needed for the structure, in amperes (A)

T_{design} = design life (period between drydocking) for the anode system, in years (y)

u = utilisation factor determined by the portion of anodic material consumed when the remaining anode material cannot deliver the current needed (dimensionless). The shape of the anode and the design of the steel core within it will affect the utilisation factor, which may be in the range of 0.7 to 0.95

Q = practical current capacity for the anode material in the environment considered, in A h/kg (A h/lb) (refer to EN 12496)

8670 = number of hours per year.

2.7.4 Anode Life

The anode life may be determined using the following formula:

$$T_{anode} = \frac{(W_{anode} \times u)}{(E \times I_s)} \quad \text{years}$$

where

T_{anode} = effective lifetime of the anode, in years (y)

T_{anode} = net mass of anode alloy (excluding the steel core), in kilograms (kg) [pounds (lb)]

u = utilisation factor determined by the portion of anodic material consumed when the remaining anode material cannot deliver the current needed (dimensionless). The shape of the anode and the design of the steel core within it will affect the utilisation factor, which may be in the range of 0.7–0.95

E = consumption rate of the anode material in the environment considered, in kg/(A-y) [lb/(A-y)]

I_s = average (mean) current output of the anode during the lifetime, in amperes (A).

2.7.5 Anode Resistance at the End of Life

Anode resistance will increase with time if consumption of the anode results in dimensional changes. The design should assess the current output capability of the anode at the end of life.

2.7.5.1 Calculation of Anode Weight at End of Anode Life

For all anode shapes:

$$W_{final} = W_{initial}(1 - u)\text{kg (lb)}$$

where

W_{final} = net mass of anode alloy (excluding the steel core), in kg (lb)
$W_{initial}$ = initial value, in kg (lb)
W_{final} = final (or end of life) value, in kg (lb)
u = utilisation factor determined by the portion of anodic material consumed when the remaining anode material cannot deliver the current needed (dimensionless). The shape of the anode and the design of the steel core within it will affect the utilisation factor, which may be in the range of 0.70 to 0.95.

2.7.5.2 Calculation of Anode Dimensions at End of Anode Life

2.7.5.2(a) **For Slender Anodes**

$$L_{final} = L_{initial} - (0.1 \times u \times L_{initial}) \quad \text{m (in)}$$

where
L = length of the anode alloy, in m (in)
$L_{initial}$ = Initial Value, in m (in)
L_{final} = Final (or end of life) Value, in m (in).

The depleted anode, with its steel core, is then assumed to be a cylinder with length L_{final} and its cross-sectional area is calculated from the estimate of W_{final} above, the density of the anode alloy, and the volume of the anode core within the final length of the anode.

$$X_{final} = \pi \frac{W_{final}}{L_{final} d_{anode}} + X_{core} \quad \text{m}^2 (\text{in}^2)$$

$$r_{final} = \sqrt{\frac{X_{final}}{\pi}} \quad \text{m(in)}$$

where

X_{final} = final (or end of life) value cross-sectional area of the anode (including the core), in m^2 (in^2)

X_{core} = cross-sectional area of the core, in m^2 (in^2)

d_{anode} = specific gravity of the anode alloy, in kg/m^3 (lb/in^3)

r_{final} = final (or end of life) anode radius, in (in).

Final anode resistance is then determined according to relevant resistance formulae using values of r_{final} and L_{final}, as appropriate.

2.7.5.2(b) For Long Flush Mounted Anodes

Calculate as for slender anodes, but assuming that the final shape is a semi-cylinder:

$$r_{final} = \sqrt{\frac{2X_{final}}{\pi}} \quad \text{m (in)}$$

2.7.5.2(c) For Short Flush or Bracelet Anodes

Assume that the resistance does not change from the Initial Value. The calculation of final resistances may be based on estimates of the final geometry and dimensions of the anode. More specific, additional suggestions are as follows.

(1) *For an Offset Type Anode.* It may be assumed that the final cross-sectional shape of the anode is cylindrical, refer to Fig. 2.2. The final radius of the anode cylinder, when it has reached the limit of the anode utilisation factor, u, or 0.90, is:

$$FinalRadius = (InitialRadius - CoreRadius)(1 - u) + CoreRadius$$

If the core radius is not known, it can conservatively be set to zero. However, this could be overly conservative.

The length of the anode should be reduced by 10% when computing the final anode resistance. However, NACE SPO176 (appendix D) does not make this adjustment.

(2) *For a Contact Anode.* The final geometry is influenced by the original shape. It may be assumed that the anode has wasted uniformly in thickness. The original thickness of the anode is unknown; conservatively, only the area of the face in contact with the structure could be considered.

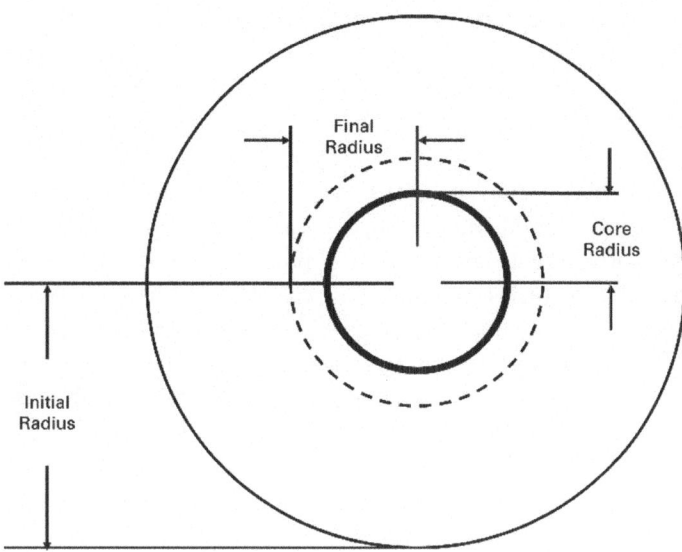

Fig. 2.2 Illustration of anode sizes

2.8   Cathodic Protection Systems

2.8.1   General

Two types of cathodic protection systems are used:

- Galvanic anodes cathodic protection system; and
- Impressed current cathodic protection system.

A combination of both systems may be used (hybrid). Galvanic anode systems employ reactive metals as auxiliary anodes that are in direct electrical connection with the steel to be protected. The difference in natural potentials between the anode and the steel, as indicated by their relative positions in the electro-chemical series, causes a positive current flow in the electrolyte from the anode to the steel. Thus, the steel surface becomes more negatively charged and becomes the cathode. The metals commonly used as sacrificial anodes are aluminium, zinc, and magnesium. These metals are alloyed to improve the long-term performance and dissolution characteristics. Chapter 3 addresses sacrificial anode systems. Impressed-current systems employ inert (very low dissolution) anodes and use an external source of DC power (rectified AC) to impress a current from an external anode onto the metallic surface to be cathodically protected. Chapter 4 addresses impressed current cathodic systems.

2.8.2   Cathodic Protection System Selection

The following information should be considered when the cathodic protection system is selected:

(1) Selection and specification of materials, equipment, and installation practices that provide for the safe installation and operation of the cathodic protection system
(2) Specification of materials and installation practices in conformance with applicable code, regulations, and Class requirements with dependable service for the design life
(3) Selection and design of the cathodic protection system for optimum economy of installation, safety, and lifetime maintenance and operation; and
(4) Selection of a cathodic protection system to minimise detrimental effects on coatings and steels to be protected by cathodic protection systems.

To help with selection of cathodic protection systems, a comparison of sacrificial and impressed current anode systems is given in Table 2.6.

2.8.2.1   Use of Galvanic Anodes
Use of galvanic anodes is appropriate under the following conditions:

- When a relatively small amount of current is needed
- Usually when lower-resistivity electrolytes, such as mud/soil and seawater, are present
- For local cathodic protection to provide current to a specific area on a structure; and
- When additional current is needed at problem areas, such as isolated points from overall impressed current cathodic protection systems or for electrically shielded areas caused by non-uniform current distribution from remotely located impressed current systems.

2.8.2.2   Use of ICCP Systems
Impressed current cathodic protection systems are most common in ships today. Where galvanic anodes are used, they are normally placed in the vessel stem, bilges, and seawater intakes. ICCP systems are typically used:

- When there are large current needs, particularly for bare or poorly coated structures
- On surface ships and submarines
- In all electrolyte resistivity; and
- To overcome stray current or cathodic interference problems.

Table 2.6 Comparison of galvanic anodes and impressed current cathodic protection systems for ship hulls

Comparison item	Sacrificial anode systems	Impressed current systems
Environment	Only practical for low resistivity of electrolyte, such as soils and waters	Less restriction from resistivity. However, the anodic reaction products from seawater may be detrimental to coatings and structure
Design and installation	Simple in design and installation, but costly installation because of intensive labor. Wrong connection is not possible	Needs careful design and installation. Wrong connection is possible. Initial cost is higher, but life cycle cost is lower
Power source	No electric power supply is needed. Electrons are provided by the sacrificial anodes. Can be used where electrical power is not available	Continuous DC power supply is required
Anodes	Bulk of anode material may restrict water flow and increase weight/turbulence/noise/drag on ship hull	Lighter and fewer in number. Anodes may be designed to have minimum effect on water flow. Low hull profile reduces noise and drag
Control	Cathodic protection current is not controllable, which is to be self-adjusting based on anode number and size. Always active to provide protection, while properly installed and immersed, until the anode is consumed	Cathodic protection current is controllable with large availability of current. System operation may be automated and can be monitored remotely. Regular maintenance and monitoring are needed
Interaction	Less likely to affect any neighbouring structures	Effects on other structures situated near the anodes need to be assessed
Damage	Where a system comprises a large number of anodes, the loss of a few anodes has little overall effect on the system	Loss of anodes can be very critical to the effectiveness of a system

(continued)

Table 2.6 (continued)

Comparison item	Sacrificial anode systems	Impressed current systems
Maintenance	Generally, no maintenance and supervision required. Anode replacement is needed when consumed about 50% or more	• Equipment designed for long life. No replacement of system components is required except for reference electrodes, which last 10 years • More technical and trained personnel are necessary for cathodic protection supervision • Regular checks and log-keeping required on electrical equipment in service • Anode replacement is not required; and • The system is frequently shut off in port, resulting in periods of no protection for the hull

2.8.3 Installation Consideration

All components of the cathodic protection system should be installed at locations where the probability of disturbance to ship operations or mechanical damage is minimal. The anode assembly and its attachment should be designed to be highly resistant to mechanical damage. Generally, when few anodes are involved for high current outputs, the loss of an anode may significantly reduce the performance of the system. Anodes should not be located in:

- Areas where they can cause problems in the normal operation of the ship
- High stress areas or areas subject to high fatigue loads; and
- Areas where they could be damaged (by craft coming alongside, anchor chains or cables).

The number, dimensions, and location of anodes should be determined in order to be able to deliver the maximum protection current demand I_{max} and to achieve the cathodic protection criteria for the entire cathodic protection zone protected by that cathodic protection system.

2.9   Electrical Continuity

2.9.1   General

In addition to ship hulls, cathodic protection is necessary for metallic appurtenances such as rudders, stabilisers, propellers, and thrusters. Electrical bonding of the metallic appurtenances to the hull should be provided by appropriate means unless the appurtenances are protected by independent cathodic protection systems. This electrical bonding with low resistance should be maintained to provide adequate cathodic protection of the appurtenances connected:

(1) To prevent galvanic corrosion of the hull or bearings, it is necessary to bond corrosion-resistant copper-based alloys or stainless-steel propellers or thrusters to the adjacent hull
(2) Rudders and stabilisers should be bonded by means of flexible cables connected to adjacent hull generally by welded/brazed studs; and
(3) Cable connections to the hull should be of a welded or brazed type. Coatings on contact surfaces should be removed prior to assembly. If the contact is made by using copper cables welded/brazed at each end, these cables should be stranded and have a minimum cross-section of 16 mm^2 (0.025 in^2). If cable shoes are used, the copper cable should be brazed to the cable shoe.

2.9.2   Shaft Ground Assembly

A turning propeller shaft on a ship is electrically insulated from the hull by the lubricating oil film in the bearings and by the use of non-metallic bearing materials in the tail shaft. When the shaft is insulated in this way, an electrical potential can be measured between the shaft and the hull, and this can cause corrosion. If the ship has a system of cathodic protection, whether it is sacrificial anode or an impressed current system, the shaft insulation will prevent the propeller and the boss from receiving cathodic protection. The current entering the propeller could exit the shaft to the seawater at the bearing components and cause accelerated corrosion of the shaft. The effectiveness of the shaft ground assembly system should provide a maximum contact resistance of no greater than 0.001 ohms for a water filled bearing and 0.01 ohms for an oil filled bearing. The potential readings through mV meter should be checked and maintained below a level. A maximum value of 40 mV is acceptable unless otherwise specified. Figure 2.3 illustrates a typical arrangement for shaft grounding and monitoring.

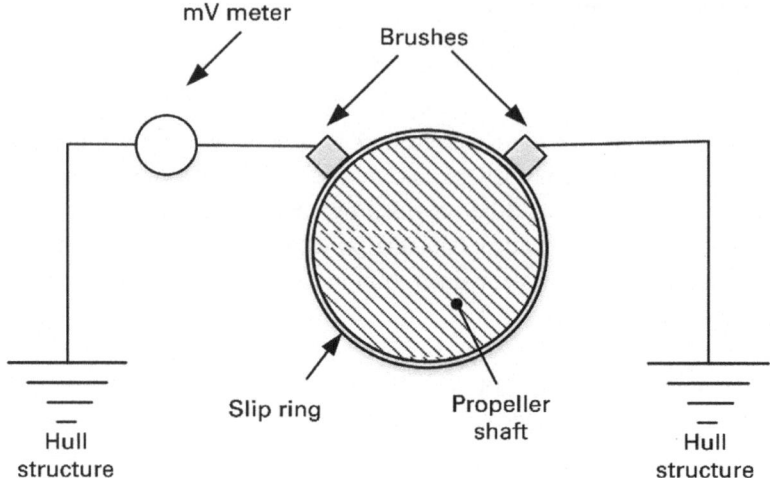

Fig. 2.3 Typical arrangement for shaft grounding and monitoring

2.9.3   Rudder (and Other Appendages) Ground

Rudder posts and stabilisers, if any, should be bonded to adjacent hull by means of flexible cables. Brazing a braided, tinned, grounding strap, at least 3.8 cm (1.5 in) wide, between the stock and the hull grounds rudders, planes, and stabilisers. Allowances should be made for rudder movement by providing a large loop of ground strap.

Galvanic Anodes System

3

3.1 General

A galvanic (also called sacrificial) anode is a metal that has a more negative potential with respect to the structure metal to be protected. Galvanic anodes corrode preferentially to the protected structure, providing protection to the structure. The objective of a galvanic anode system is to deliver sufficient current to protect part of the structure for the designed life of the system. There are three basic components of a marine galvanic anode cathodic protection system:

(1) Anode(s)
(2) Connecting System, including cabling and fasteners; and
(3) Protected Structure.

Sacrificial anode materials are usually alloys of aluminium, zinc, or magnesium.

3.2 Design Considerations

A detailed design of galvanic anode systems should, as a minimum, include the following:

(1) Acceptance criteria for the completed system
(2) Detailed drawings and specifications of anode materials, sizes, and attachment
(3) Detailed calculations with specified design current density and anode resistance
(4) Detailed specification for inserts, attachment, and anode/structure continuity; and
(5) Detailed specification for installation, testing, commissioning, and operation.

© The Author(s), under exclusive license to Springer Nature Switzerland AG 2025
A. A. Olsen, *Cathodic Protection of Marine Vessels*, Synthesis Lectures on Ocean
Systems Engineering, https://doi.org/10.1007/978-3-031-77578-9_3

3.3   Electrochemical Properties of Anodes

3.3.1   General

The anode material's electrochemical properties may include potential, current capacity, and anode consumption rate. The performance of a galvanic anode material (alloy) is dependent on its actual chemical composition and homogeneity, current density, and the environmental conditions it is exposed to. In addition, anode surface morphology can affect the efficiency. The properties of an anode alloy may be obtained from the performance data in the given environmental conditions. The information that should be documented includes:

- Driving voltage (i.e., the difference between closed circuit anode potential and the positive limit of the protection potential criterion)
- Practical electrical current capacity [A-h/kg (A-h/lb)] or consumption rate [kg/A-y (lb/ A-y)]
- Susceptibility to passivation; and
- Susceptibility to intergranular corrosion.

Where long-term performance data (at least 12 months duration) is not available for a specific alloy/environmental combination, the properties of the anode material should be determined by appropriate tests, and caution should be exercised in selecting values for design purposes. For details of anode design and acceptance criteria, refer to EN 12496 Section 5. More information can also be found in NACE SP0492 and NACE SP0387.

3.3.2   Material Chemical Composition

Galvanic anode material performance is related to the chemical composition. Therefore, strict control of the alloy chemical composition of both the alloying elements and impurities is essential. Electrochemical testing should be carried out to confirm the relevant anode operating potential and anode capacity for a particular environment when there are no reliable historical or laboratory data available for a specific anode alloy. The same kind of testing is also to be performed on any field-proven but not qualified alloy composition by a supplier. The laboratory test procedure should be selected to provide best representation of the expected operating conditions (including electrolyte, temperature, and anode current density).

3.3.3 Potential

The selected anode alloy is to have a closed-circuit potential more negative than the protection potential needed for cathodic protection. The operating potential of the selected anode alloy should be stable with time to provide long-term performance for the particular operating environment, which should be documented by long-term testing. Anode alloys will polarise when current is applied. The potential will also vary with the surrounding environment, surface condition (corrosion products forming and removing on the anode surface), and variations in the current demand. The anode operating potential is generally more negative than -1.00 V measured with an Ag/AgCl/seawater reference electrode. However, where a low driving voltage is necessary, either special anode compositions (with an operating potential of -0.85 V versus Ag/AgCl/seawater reference electrode) or anodes with a voltage controller (such as diode or resistive bond) between the anode and the structure can be used.

3.3.4 Current Capacity

The current capacity for a galvanic anode alloy is the total amount of electricity (A-h) produced by one kilogram (one pound) of the anode material for a given operating condition. The practical current capacity, which is used in cathodic protection design calculation, is lower than the theoretical current capacity due to anode's self-corrosion. The current capacity of galvanic anodes is temperature dependent and should be given for protected components at elevated temperature. The practical current capacity decreases with increasing temperature both in seawater and in marine sediments. The practical current capacity is also anode current density dependent. It increases with increasing anode current density. At extremely low current densities, an anode's self-corrosion may be more pronounced and can give a significant reduction in the practical current capacity.

In addition to the anode potential, the practical current capacity of an anode alloy should be documented by long-term testing for the particular operating environment. Where long-term performance data relating to anode capacity is not available for a specific alloy/environmental combination, additional tests should be carried out to determine the effect of current density, temperature, and time on the current capacity of the alloy for the particular environment and the various operating conditions of the anode.

3.3.5 Anode Consumption Rate

The anode consumption rate for a galvanic alloy anode is the total amount of anode material consumed for a current output of one ampere for one year (kg/(A-y) [lb/(A-y)]). Like the current capacity, a practical consumption rate is higher than the theoretical consumption rate. The anode consumption rate and the current capacity are related by:

$$E \times Q = 8760 \quad \text{h/y}$$

where

$E =$ anode consumption rate, in kg/(A-y) [lb/(A-y)]
$Q =$ current capacity, A-h/kg (A-h/lb)
$8760 =$ number of hours in one year.

3.4   Anode Materials and Their Properties

3.4.1   General

Most common galvanic anodes used are aluminium (Al), magnesium (Mg), or zinc (Zn) based alloys. The pure forms of these metals are often not suitable as anodes because they undergo too much self-corrosion and do not stay active. Thus, alloyed aluminium, magnesium, and zinc anodes are formulated to minimise this effect. Galvanic anodes for use on ship hulls are usually made of zinc- or aluminium-based alloys. Magnesium based alloy anodes are suitable for freshwater applications, but not for seawater application because of their high potential and high conductivity of seawater. The chemical composition of any alloy used for galvanic anodes should be specified by the supplier and the corresponding electrochemical properties should be tested and documented.

3.4.2   Zinc Anodes

Zinc anodes require the use of high-quality zinc with the maximum allowable trace element of iron. The addition of cadmium and aluminium will permit an increase in the tolerance for iron. Zinc anodes have a natural potential approximately -1030 mV with reference to a silver/silver chloride reference electrode in seawater. Temperature limitations for zinc alloy anodes are a concern [max. 60 °C (140 °F)] and PH value range (5–10). The typical composition of zinc anodes is given in Table 3.1. Electrochemical properties of zinc anodes in seawater are listed in Table 3.2.

3.4.3   Aluminium Anodes

Aluminium is preferred for seawater applications because it has a much lower consumption rate than magnesium or zinc. Aluminium alloys are not normally effective in fresh water. Environments such as brines, seawater of varying salinity, marine sediments, or hydrogen sulphide (H_2S), result in performance variations of aluminium anode alloys.

Table 3.1 Chemical composition of zinc anodes, weight %

Element	EN 12496				ASTM B418		MILA-18001K
	Alloy Z1 (MIL-A-18001K or ASTM B418 Type I)	Alloy Z2 (ASTM B418 Type II)	Alloy Z3, Limited to 50 °C (122 °F)	Alloy Z4, up to 60–80 °C (140–176 °F)	Type I Limited to 50 °C (122 °F)	Type II	Limited to 50 °C (122 °F)
Al	0.1–0.5	0.005 max	0.1–0.2	0.1–0.25	0.1–0.4	0.005 max	0.1–0.5
Cd	0.025–0.07	0.003 max	0.04–0.06	0.001 max	0.03–0.1	0.003 max	0.025–0.07
Fe	0.005 max	0.0014 max	0.0014 max	0.002 max	0.005 max	0.0014 max 0.005 max	
Cu	0.005 max	0.002 max	0.005 max	0.001 max	–	–	0.005 max
Pb	0.006 max	0.003 max	0.006 max	0.006 max	0.03 max	0.003 max	0.006 max
Mg	–	–	0.5 max	0.05–0.15	–	–	–
Sn	–	–	0.01 max	–	0.001 max	0.001 max	–
Others	0.1 max	0.005 max	0.1 max	0.1 max	–	–	–
Zn	Remainder	99.99 min	Remainder	Remainder	Remainder	99.99 min	Remainder

Table 3.2 Typical electrochemical properties of zinc anodes in seawater

Alloy type	Operation temperature °C (°F)	Closed circuit potential to Ag/ AgCl/Seawater reference electrode, V	Practical current capacity A-h/kg (A-h/lb)	Practical anode consumption rate kg/A-y (lb/A-y)
Alloy Z 1 or MIL-A-18001K or ASTM B418 Type I	5–25 (41–77)	−1.03	780 (354)	11.2 (24.7)
Alloy Z 2 or ASTM B418 Type II	5–25 (41–77)	−1.00	760 (345)	11.5 (25.4)
Alloy Z 3	5–25 (41–77)	−1.03	780 (354)	11.2 (24.7)
Alloy Z 4	5–25 (41–77)	−1.03	780 (354)	11.2 (24.7)
	60–80 (140–176)	−0.97	690 (313)	12.7 (28.0)

Note The current capacity and anode consumption rate values shown are the practical values for zinc alloys, which incorporate an allowance for self-corrosion of the alloy already. No further efficiency allowance is necessary

Anode capacity for all aluminium-based alloys will be significantly reduced at elevated operating temperatures. For example, anode capacity for alloys A1 and A2 will drop essentially linearly from 2500 A-h/kg (1134 A h/lb) at 25 °C (77 °F) to 500 A-h/kg (227 A-h/lb) at 80 °C (176 °F). This is equivalent to a change in anode consumption rate of 3.5 kg/(A-y) [7.7 lb/(A-y)] at 25 °C (77 °F) to 17.5 kg/(A-y) [38.6 lb/(A-y)] at 80 °C (176 °F). Typical composition of aluminium anodes is given in Table 3.3. Electrochemical properties of the aluminium anodes in seawater are listed in Table 3.4.

3.4.4 Magnesium Anodes

Magnesium anodes are rarely used in seawater because of the potential for overprotection and low anode efficiency. Magnesium anodes are used in fresh water and in specialised situations. Magnesium anodes are not used on aluminium hulls because aluminium is an amphoteric material. The typical composition of magnesium alloy anodes is given in Table 3.5. Electrochemical properties of magnesium alloy anodes in seawater are listed in Table 3.6.

Table 3.3 Chemical composition of aluminium anodes, weight %

Element	EN 12496				ASTM B418	MIL-DTL-24779B	
	Alloy A1 for Marine Applications	Alloy A2 for Offshore Applications	Alloy A3 for Deepwater and Cold-Water Applications	Alloy A4 for Low Driving Voltage Applications		Conventional	Low Voltage
Zn	2-6	3-5.5	4.75-5.75	0.15	max	5-6 4-6.5	< 0.15
In	0.01-0.037	0.016-0.04	0.016-0.02	0.005	max	---	0.014-0.02
Ga	---	---	---	0.092-0.11	---	---	0.092-0.11
Fe	0.12 max	0.09 max	0.06 max	0.08 max	0.17 max	0.09 max	< 0.08
Cu	0.006 max	0.005 max	0.003 max	0.005 max	0.02 max	0.004 max	< 0.005
Si	0.12 max	0.1 max	0.08-0.12	0.1 max	0.1 max	0.08-0.2	< 0.1
Cd	0.002 max	0.002 max	0.002 max	---	---	---	---
Mercury Hg						0.001 max	< 0.005
Tin Sn						0.001 max	0.001 max
Ni						---	< 0.005
Magnesium Mg						---	< 0.01
Manganese Mn						---	< 0.01
Others Impurities (each)	0.02 max	0.02 max	0.02 max	0.02 max	---	0.020	---
Others Impurities (total)	0.1 max	0.1 max	0.1 max	0.1 max	---	0.10	---
Al	remainder	remainder	remainder	remainder	remainder	remainder	min 99.8

Table 3.4 Typical properties of aluminium anodes in normal operation conditions (anode current density, temperature, and seawater)

Alloy type	Closed circuit potential Ag/AgCl/ Seawater reference electrode, V	Practical current capacity A-h/kg (A-h/lb)	Practical anode consumption rate kg/ (A-y) [lb/(A-y)]
Alloy A1	1–1.09	2500 (1134)	3.5 (7.7)
Alloy A2	1–1.09	2500 (1134)	3.5 (7.7)
Alloy A3	1–1.09	2500 (1134)	3.5 (7.7)
Alloy A4	−0.83	1500 (680)	5.8 (12.8)
MIL-DTL-24779B conventional	−1.05 to −1.15 long term −1.05 to −1.15 short term	2535 (1150) long term 2535 (1150) short term	
MIL-DTL-24779B low voltage	−0.80 to −0.90 long term −0.78 to −0.83 short term	1656 (751) long term 1800 (816) short term	

Note The current capacity and anode consumption rate values shown are the practical values for aluminium alloys, which incorporate an allowance for self-corrosion of the alloy already. No further efficiency allowance is necessary

Table 3.5 Chemical composition for magnesium alloy anodes, weight %

Element	EN 12496		MIL-A-21412
	Alloy M1 (ASTM B843-93)	Alloy M^2	
Mn	0.15–0.7	0.5–1.5	0.15 min
Al	5–7	0.05 max	5–7
Zn	2–4	0.03 max	2–4
Fe	0.005 max	0.03 max	0.003 max
Cu	0.08 max	0.02 max	0.1 max
Si	0.3 max	0.05 max	0.3 max
Pb	0.03 max	0.01 max	NA
Ni	0.003 max	0.002 max	0.003 max
Others	Total 0.3 max	Each 0.05 max	Total 0.3 max
Mg	Remainder	Remainder	Remainder

Table 3.6 Typical electrochemical properties of magnesium anodes at ambient temperatures 5 °C–25 °C (41 °F–77 °F)

Alloy type	Environment	Closed circuit potential to Ag/AgCl/Seawater reference electrode, V	Practical current capacity A-h/kg (A-h/lb)	Practical anode consumption rate kg/(A-y) [lb/(A-y)]
Alloy M1 or ASTM B843	Seawater	−1.50	1200 (544)	7.3
Alloy M^2	Seawater	−1.70	1200 (544)	7.3 (16.1)

3.5 Anode Arrangement

3.5.1 Anode Connection

Anodes, including cores and supports, should be designed so as to meet specified performance requirements during fabrication, transport, installation, and operation. The dimensions and shape of the anodes, steel core, and attachments should be designed to withstand the mechanical forces that may act on the anodes. Sacrificial anodes are usually cast and contain inserts of a less electronegative material (e.g., steel) to maintain electrical continuity and mechanical strength towards the end of the anode life. The inserts should be prepared prior to casting to provide maximum electrical contact with the anode alloy. For all anodes, the anode and anode core dimensions should be designed for the proposed fitting requirements. Anode cores should be fabricated from weldable structural steel plate/sections and be compatible with the steel of the structure or structural elements to which they are attached. Galvanic anodes can be attached to the structure either directly, by welding or bolting integral inserts to the structure (e.g., hull mounted anodes), or by a wire connection between the anode and structure. Galvanic anodes may be applied directly to the hulls without shielding. However, better current distribution is obtained with a dielectric shield between the anodes and the hull metal, if the anode is electrically connected to the hull metal being protected.

A low electrical resistance contact between the anode and the hull is to be maintained throughout the operating life of the anode. When the anode's steel inserts are attached to the ship hull by welding, stresses should be minimised at the weld location. The steel insert may be bolted to supports which have been welded to the structure. An attachment to studs "fired" into the structure is not permitted. Stud welding may be permitted. The welding of the anodes or of the supports should be performed in accordance with class requirements. If a wire is used, the wire connection to the anode should be performed by the manufacturer and multiple cables should be utilised to provide a degree of redundancy in case of cable damage. The cable should be multi-stranded with a minimum cable size of 10 mm^2 (0.016 in^2) cross-sectional area. The cable lug should be connected to the anode core by either a mechanical (i.e., bolted) connection or direct welding. The contact point should be protected using a suitable coating (epoxy or equivalent). Cable attachment to the structure should provide both a mechanically secure and a low electrical resistance connection. Pin brazing, thermite welding, or "volcano" bolts or lock nuts with serrated washers as appropriate can be used. The wire should be coated with a dielectric insulation, and the connections should be coated.

Sacrificial anodes on a shaft should be installed so as not to throw the shaft out of balance or restrict water flow to strut bearings. Collars applied to the propeller shaft are usually adequate to protect the propellers and shafts of bronze and stainless steel. In the case of thin shell plates or sensitive materials, a welded doubler plate with sufficient thickness and an extra border of 20 mm (0.79 in) on all sides around the welding points of the anode is needed. Where anodes are to be attached to the hull plating of fuel oil or oil cargo tanks or internally coated areas, the use of doubler plates should be addressed. If used, they should be continuously welded to the hull. The anodes should not be attached to bottom plating or large unsupported panels and preferably not directly to the shell plating within the major longitudinal strength sections.

3.5.2   Anode Distribution on Hull

Anodes should be mounted on surfaces in such a way to avoid entrapping gas bubbles and disturbing water flow near intakes and discharge fittings. Anodes should be located so as not to disturb the flow of water past the propeller(s) or jet drive intake and nozzle(s). On the ship's hull, flat anodes should be specified to minimise the flow resistance.

The anodes should be uniformly distributed over the underwater surfaces to achieve good current distribution so as to protect the entire ship. It may be advantageous in certain situations to use computer modelling based on finite elements or boundary elements calculation methods and/or model testing. They should not be attached to areas likely to sustain regular mechanical damage (e.g., in way of the anchor). In way of the bilge, the anodes should be arranged so that they cannot be damaged when the ship is berthed. Where a bilge keel is fitted, the anodes should preferably be alternately attached on the upper and lower sides. If the bilge keel height is not sufficient for this, the anodes should be arranged on the hull near the bilge keel and alternate between above and below the bilge keel. The spacing of the anodes on the bilge in the middle of the ship should not be greater than 6–8 m (19.7–26.2 ft) to provide an overlap of the protected zones. For ships sailing in water with low resistivity (e.g., the Baltic) or requiring high protection current densities (e.g., the tropics), the spacing range is smaller [such as 5 m (16.4 ft)]. It is reduced even further for ships subjected to a large amount of mechanical damage (e.g., ice-going ships in arctic waters).

A distinction is made between complete and partial protection of the underwater area, depending on the extent of the protected region. Complete protection of the ship with galvanic anodes or impressed current is of increasing importance since defects in the coating due to mechanical damage are more frequent, particularly at the bow and amidships. In the case of partial protection, only the stern is protected, because of the high flow rate and aeration, as well as the formation of corrosion cells on attachments, such as the propeller and rudder. Partial protection can also be extended to the bow, which also experiences high rates of flow. The most efficient protection is achieved by the use of a large number of small anodes well distributed around the hull. For ships where only the stern is protected, 25% to 30% of the total anode weight for complete protection is needed. With this partial protection of the ship, at least two anodes of the same shape, or 10% of the actual stern protection, should in addition be applied at 3–8 m (9.8–26.2 ft) in front of the front stern anode. The remaining anodes should be distributed evenly on the bow and amidships, preferably along suitable flow lines in order to minimise drag.

For special hull types (e.g., hydrofoils, ships with waterjet drives, catamarans), the structural design and the flow rate should be considered for the arrangement of the external protection. If there is no central cathodic protection system, rudders should be cathodically protected by anodes, and propellers and shafts by zinc rings affixed to the propeller hubs or shafts.

3.5.3 Anodes Near the Bow

The anodes near the bow should be arranged in the direction of water flow and placed so that they cannot be damaged by the anchor chain. Because of the high loading, anodes should be installed not only on the bilge but also in the vicinity of the flat plate keel.

3.5.4 Anodes at the Stern

At the stern, the anodes should be predominantly located in the area of the stern tube, sole piece, propeller well, and in some cases, the heel. Care should be taken that eddies caused by the anodes do not impinge on the propeller. For this reason, no anodes should be installed in the 0.4D to 1.1D forbidden region (D is diameter of the propeller). It has also been required that the anodes in the area of the stern tube be at least a distance of D from the propeller (refer to Fig. 3.1). The anodes attached over the propeller well are sometimes not immersed when the ship is at a shallow draft, so the anodes should be fixed obliquely on the stern profile. This principle also applies to anodes above the forbidden region. Above the propeller well and the heel piece just before the propeller well, at least one anode should be mounted on each side. In way of the stern tube exit, the necessary anodes should be arranged so to give special attention to the anode-free area, with at least one on each side (Fig. 3.1).

To protect the shaft brackets, anodes should be applied near their mountings on both sides of the hull. The size and material of the shaft brackets should be taken into account

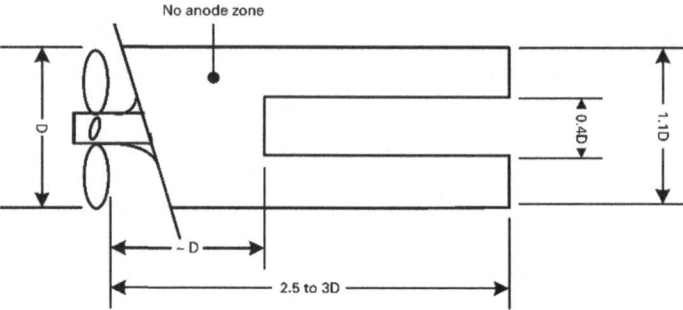

Fig. 3.1 Arrangement of galvanic anodes on the stern

in determining the number of anodes. Propeller brackets in multi-propeller ships should be particularly protected. On small ships, the anodes should be attached on both sides at the base of the propeller bracket. On large ships, the anodes should be welded onto the propeller brackets.

3.5.5 Protection for Propellers and Shafts

As a rule, propellers, shafts, and struts should also be included in the cathodic corrosion protection of the outer shell. These parts should be connected conductively with the hull by means of slip rings on the propeller shafts and brushes. Refer to refer to Chap. 2, Sect. 2.9.2 for guidance on shaft ground assembly. It is possible to cathodically protect the propeller and shaft solely through a zinc ring mounted on the propeller hub or on the shaft. Relevant calculations and distribution of the anodes are also necessary for special propulsion. For Kort nozzles, a basic protection current density of 25 mA/m^2 (2.3 mA/ft^2) is used for the total surface area. The anodes should be attached on the external surface at a spacing of 0.1–0.25r at the region of greatest diameter. Internally, the anodes should be fixed to the strengthening struts. With Voith-Schneider propellers, the anodes are arranged around the edge of the base of the propeller.

3.5.6 Anodes for Rudders

The rudder should be provided with anodes on both sides. These should be fixed either at the level of the propeller hub or as far as possible above and below the rudder blade. There are anodes specially designed for use on rudders, which are welded to the front edge of the rudder. Rudder heels should be given one anode on either side. The width of the anode should be smaller than the height of the rudder heel. The rudders of high-speed craft (capable of speeds over 30 knots) should as a rule only be protected by anodes adapted to the rudder profile. If this is not possible, the rudder should be included in the complete protection scheme by cable or copper-band connections to the hull.

3.5.7 Anodes for Sea Chests and Scoop Openings

The anodes should be arranged so that a shadow effect is avoided as much as possible. Openings in the outer shell, such as sea cheats, scoop openings, lateral thrust propellers, or similar, should be protected by externally placed anodes only up to a depth of one to two times the opening diameter. Where anodes are to be fitted for the protection of bow thruster units, they should be fitted as close as possible to the vulnerable areas, but with due consideration to the water flow. Where anodes are to be fitted within sea chests, consideration should be given to stray current interference to other items included in the sea chest.

3.5.8 Cathodic Protection of Heat Exchangers, Condensers, and Tubing

Galvanic or impressed current anodes are used to protect heat exchangers, condensers, and tubing components. The appropriate anode material is determined by the electrolyte (zinc and aluminium for seawater and magnesium for freshwater). Platinised titanium should be used for the anode material in impressed current protection. Potential-regulating systems working independently of each other should be used for the inlet and outlet feeds of heat exchangers due to the different temperature behaviour. The protection current densities depend on the material and the medium. The internal cathodic protection of pipes is only economically feasible for pipes with an internal diameter greater than 400 mm (15.75 in) due to the limit on range. Internal protection can be achieved in individual cases by inserting local platinised titanium wire anodes.

3.6 Cathodic Protection of Internal Tanks

3.6.1 General

This Subsection applies for the cathodic protection of the internal areas of ships by means of sacrificial anodes. Cathodic protection in closed compartments without ventilation may cause development of hydrogen gas to an extent that an explosive gas mixture (i.e., hydrogen/oxygen) may eventually develop. The risk is moderate with aluminium- and zinc-based galvanic anodes. For ballast tanks and other tanks with seawater, only a galvanic anode cathodic protection system should be applied. The impressed current system

Table 3.7 Recommended initial, final and mean design current densities for bared steel surface exposed to seawater with a range of surface water temperatures

Surface water temperature, °C (°F)	Initial mA/m^2 (mA/ft^2)	Final mA/m^2 (mA/ft^2)	Mean mA/m^2 (mA/ft^2)
>20 (68)	120 (11.1)	80 (7.4)	60 (5.6)
12–20 (54–68)	140 (13.0)	90 (8.4)	70 (6.5)
7–11 (44–52)	170 (15.8)	110 (10.2)	80 (7.5)
<7 (44)	200 (18.6)	130 (12.1)	100 (9.3)

may generate excessive hydrogen gas, which may be hazardous. Cathodic protection is necessary for freely flooded compartments and for closed compartments with free access to air. Closed and sealed flooded compartments do not normally need cathodic protection. The protective duration should be at least 5 years (43,800 h) or defined by agreement with the shipowner. This Subsection applies only for internal surfaces which are exposed to an electrolyte with sufficient conductivity for at least 50% of exposure time. The effectiveness of the anodes is limited in fresh water and river water. Uncoated stainless steels are not protected cathodically if they are suitable for the corrosion load. However, coated stainless steels should be cathodically protected.

3.6.2 Protective Current Density

For trim tanks, ballast water tanks and cells, tank tops (inner bottoms), bilge water collector cells, slop, and sludge tanks, or similar, the following current densities are stipulated for steel:

- Coated surfaces: 10 mA/m^2 (0.9 mA/ft^2); and
- Uncoated surfaces: 120 mA/m^2 (11.1 mA/ft^2).

Depending on the load, coating, and accessibility, the protective current densities can be between 10 mA/m^2 (0.9 mA/ft^2) and 120 mA/m^2 (11.1 mA/ft^2). For freely flooded compartments and for closed compartments with free access to air, design current densities are recommended in Table 3.7.

3.6.3 Anode Selection and Weight

Anode materials should be selected in accordance with this chapter, Sect. 3.4. Magnesium or their alloys are not acceptable anodes in tanks, except in freshwater tanks that are not adjacent to cargo tanks. Aluminium anodes can be used for liquid cargo with flash points below 60 °C (140 °F) provided aluminium anodes are located so a kinetic energy of not more than 275 J (203 ft-lb) is developed in the event of their loosening and becoming detached. This limitation does not apply for ballast water tanks that are not adjacent to cargo tanks. This can refer to the requirement relating to "Corrosion Protection" of the Class Rules for the construction and classification of marine vessels. There is no height restriction for zinc anodes. The anode weight per zone can be calculated by:

$$M_{zone} = I_{zone} \times t_s \times \frac{f_B}{Q_g} \quad \text{kg (lb)}$$

where

I_{zone} = total protective current = $I_{zone} \times I_{zone}$ (A)

A_{zone} = area of a cathodic protection zone, in m^2 (ft^2). The maximum surface area covered by the electrolytic solution is used for the calculation

i_{zone} = necessary protective current density for a cathodic protection zone, in A/m^2 (A/ft^2)

t_s = protective period. The protective duration should be set to 5 years ($8760 \times 5 = 43{,}800$ h) or defined by agreement with the client

f_B = loading (utilisation) factor, which depends on the period in which the surface is covered with the electrolytic solution. In the case of constant loading (filled tanks/cells), the factor is to be set to 1

Q_g = electrochemical efficiency of the anode alloy, in A-h/kg (A-h/lb).

3.6.4 Arrangement of Anodes

Anodes should be arranged so that a shadow effect is avoided, even in areas with a complex structure. The arrangement is also to consider the prevention of fire and explosion. Because of uncertain or various water levels, anodes are primarily to be placed in the lower part of a tank, where the surface is most likely wet. It should be noted that several smaller anodes provide a better current distribution than one large anode of the same weight. It may be necessary to increase the number of anodes for the internal spaces for the following reasons:

- The effective zone of the anodes for low water levels
- A shadow effect caused by complex internal structures; and
- Galvanic corrosion from use of noble materials, which should be compensated locally.

It may also be necessary to provide extra anodes in addition to the total anode weight calculated according to the equation above in order to achieve the protective current needed. Aluminium anodes should be located in such a way that they are protected from falling objects. They should not be located under tank hatches or butterworth openings unless protected by adjacent structure.

3.6.5 Anode Attachment

Anodes should be attached to stiffeners or aligned in way of stiffeners on bulkhead plating, but they should not be attached to the shell. The two ends of the anode should not be attached to separate members which are capable of relative movement. Where cores or supports are welded to local support members or primary support members, they should be kept clear of end supports, toes of brackets, and similar stress raisers. Where they are welded to asymmetrical members, the welding should be at least 25 mm (1 in) away from the edge of the web. In the case of stiffeners or girders with symmetrical face plates, the connection may be made to the web or to the centerline of the face plate, but well clear of the free edges. Generally, anodes should not be fitted to a face plate of a higher strength steel.

The steel insert of galvanic anodes can be bolted to separate supports (brackets) connected to the stiffeners by continuous welding. Anodes can also be attached to flanged stiffeners by the use of bolted clamps of an approved design (Fig. 3.2). In such cases, the clamping bolts should be fitted with additional locking nuts. Electrical continuity checks should be performed, and the resistance should be such that the ohmic drop across any bolted connection is less than 10% of the design driving voltage between anode and steel structure, and in no case higher than 0.1 Ω. Tanks in which anodes are installed are to have sufficient holes for the circulation of air to prevent gas from collecting in pockets.

3.6.6 Coating

A tank's coating qualification and installation should meet requirements from the coating producer's recommendation and/or regulatory/class requirements. The coating should resist cathodic disbonding and pass test criteria, if any.

3.6.7 Lay-Up Period

Ballast tanks should be maintained either full or empty. During a long lay-up, ballast tanks should be protected by means of sacrificial anodes. Depending on the owner's inspection, maintenance, and repair (IMR) strategy, these tanks may be available for easy and cost-effective retrofitting of sacrificial anodes.

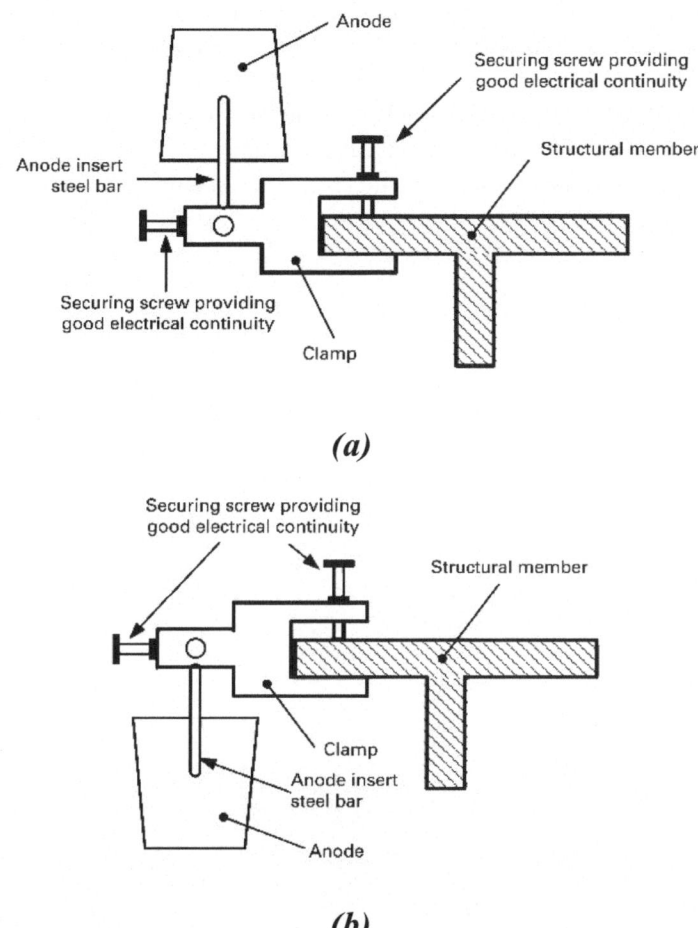

Fig. 3.2 Example of installation of a galvanic anode in a tank using clamps

3.7   Installation of Galvanic Anodes

Anodes should be inspected to confirm that dimensions and weight meet design specifications, and that any damage from handling does not affect application. In the case of
weld-on anodes, the steel cores should be inspected for conformance to specifications.
If the anode cores have welded joints or connections, these should be inspected to verify compliance with structure welding specification. Suspended galvanic anodes should be
installed after the ship is launched. The anode cables should be tested for strength, voltage
drop, and electrical contact to the structure after installation. When separate suspension
cables are used, care should be taken so that that anode lead wires are not in such tension

as to damage the lead wires or connections. The use of doubler or gusset plates should be considered for anodes mounted on standoff pipe cores and particularly for those weighing more than 230 kg (500 lb). Doubler or gusset plates may be installed on anode supports at the time of anode installation. If installed as part of the anode fabrication, these plates may be subject to serious damage during anode hauling and handling. If coatings are specified for anode supports or suspension cables, they should be visually inspected and repaired if damaged.

Impressed Current System

4

4.1   General

Impressed current cathodic protection systems (ICCP) predominate in ships today, especially when current needs and electrolyte resistivity are high. They can be automatically controlled when a ship is operating in varying conditions. The necessary protection current is automatically adjusted for changes in ship speeds and electrolyte resistivity (as when a ship enters brackish or fresh water) so that the ship is neither under protected nor overprotected. Unlike galvanic anode systems where the natural potential difference between the anode and cathode provides the driving force for current, an ICCP system is supplied with power from an external source. The power supply converts shipboard AC power to low-voltage DC and allows an adjustment of electrical current delivered to the anodes. The electrical current output delivered by the DC power source (normally a transformer rectifier) is controlled during the lifetime of the cathodic protection system in order to obtain and maintain an adequate electrochemical potential level for protection over the whole ship's hull. There are four basic components of the impressed current cathodic system:

- Anodes
- Power source
- Connecting system, including cables, connections, bonding; and
- Protected structure.

Additionally, the following components may be present:

- Monitoring and control systems
- Dielectric shields

© The Author(s), under exclusive license to Springer Nature Switzerland AG 2025
A. A. Olsen, *Cathodic Protection of Marine Vessels*, Synthesis Lectures on Ocean Systems Engineering, https://doi.org/10.1007/978-3-031-77578-9_4

- Reference electrodes
- Stuffing tubes
- Shaft ground assembly; and
- Rudder ground.

4.2 ICCP System Components

4.2.1 Transformer Rectifier Power Source, Monitoring, and Control Systems

Impressed current cathodic protection systems for ships usually include one or more transformer rectifier(s) along with multiple anodes and reference electrodes. When applicable, each cathodic protection zone (refer to Chap. 2, Sect. 2.4.2) should be protected by a dedicated system. Specific areas presenting particular situations may require the consideration of a multi-zone control system in order to adapt and optimise the electrical current distribution to the cathodic protection demand. The anode distribution should provide an even current distribution around the hull during the cathodic protection design life. The transformer rectifier should be able to deliver the necessary protective current to the anodes and to the cathodic protection zone. The maximum protection current demand (I_{max}) for a cathodic protection zone should be calculated using formulae according to the current demand (refer to Chap. 2, Sect. 2.4.5), based on the most severe service conditions. For fewer higher-current anodes, due to a less efficient electrical current distribution, the current capacity (It) of the cathodic protection system should be designed to be able to provide at least 25% more than the calculated current demand, I_{max} (i.e., $I_t \geq 1.25 I_{max}$). The transformer rectifier output voltage should take into account the voltage drops from the resistance of the electric circuit and the recommended operating voltage of the anodes. Transformer rectifiers with automatic potential control are generally used to meet the current demand necessary to maintain structure polarisation because the environment conditions and the coating condition are frequently various.

4.2.1.1 Monitoring and Control Equipment
The transformer rectifier delivers the protective current to the anodes and should be equipped with the following minimal monitoring and control equipment:

(1) A voltmeter and an ammeter for showing the DC output voltage and current
(2) A device allowing the measurement of the electrical current output of each anode
(3) Protection devices against over-voltages and short-circuits. The ability to limit the output current and anode voltage to a maximum value should be provided

(4) In the event of wire break or short circuit at the control electrodes, the protective current should be switched off automatically or regulated down to zero when in automatic mode. A locking arrangement for setting the required potential should be provided

(5) The equipment should be designed to be switched off manually if necessary

(6) A switch allowing either automatic or manual operation

(7) A monitoring system allowing the measurement of potentials with each of the reference electrodes individually, the selection of controlling reference electrode(s), and controlling electrolyte-to-structure potential independent for port and starboard sides within the protection limits to prevent over and under polarisations of the hull or appurtenance by the amount of current supplied to the anodes. The current amount is a function of coating condition, resistivity of the water, hull speed, and dock condition (whether the hull is isolated or in electrical contact with other metal structures or ships).

(8) The control precision of the set voltage for the control electrodes (target value) should be within ± 10 mV during automatic operation; and

(9) A remote warning indicating that one of these parameters is out of limit may be provided at the ship control centre to show that a system malfunction has occurred.

4.2.1.2 Indicators
The following indicators should be provided as a minimum:

(1) Indicator light "On"

(2) Indicator light "Manual Operation"

(3) Common indicator light "Malfunction"

(4) Indicator "Anode Failure or Anode Group Failure"

(5) Measurement units for "Anode Current", "Anode Voltage" and "Potential" (input impedance of the measurement circuit: ≥ 1 MΩ)

(6) Provisions for automatic or manual recording of system parameters, including output voltage, total output current, individual anode current, and hull or appurtenance/reference electrode potential should be provided. The controller may also have the capability of data logging, storage, and self-diagnosis of system components; and

(7) The AC ripple should be limited to 100 mV RMS (Root Mean Square) in order to minimise impact on the wear rate of platinum coated anodes.

4.2.2 Anodes

Unlike galvanic anodes, impressed current anodes are designed to be resistant to corrosion. Desirable properties include low resistance to current flow, physical toughness, low rate of consumption, and low cost of production. Platinum is an ideal candidate for impressed current anodes because it has a very low consumption rate, and thus longer

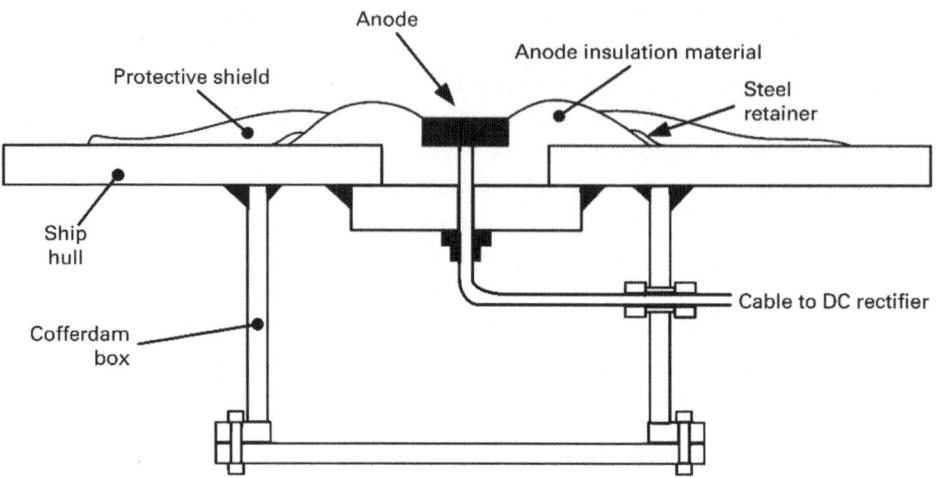

Fig. 4.1 Externally attached impressed current anode with cofferdam

service life, normally greater than drydocking intervals. It is cost effective to use platinum-coated titanium, tantalum, or niobium rather than solid platinum for the anodes. The mixed metal oxides are also usually used as the inert anodes. The anodes used in the ICCP system are usually constructed of a single continuous platinum-coated tantalum or niobium wire rod woven through an insulating glass-reinforced polyester holder. Each anode has a single hub which is bolted on the outside of the ship hull.

The anodes should be installed in sets along the hull. An anode set is composed of two anodes of similar type located on opposite sides of the hull at the same frame number and distance above the baseline. In general, they are located at the same depth, which should be greater than 1.5 m (5 ft) below the light-load line, in areas of minimum turbulence, protected from mechanical damage, and at least 4.5 m (15 ft) from either a system intake or discharge. They are recommended to be located at half minimum light ballast draft. For high-duty ships such as icebreakers, the anodes attached to the hull and their arrangements should consider the heavy mechanical wear from ice. The common anode material for ice-going vessels is platinised titanium. Recessed anodes are mainly used, and usually additional guards are provided for protection. Figure 4.1 illustrates an externally attached impressed current anode with cofferdam, and Fig. 4.2 illustrates a typical recessed current anode with cofferdam.

4.2.2.1   Platinum

Platinum has high conductivity and low consumption rate. It is used as an excellent anode material. Due to its high cost, platinum is made practical for use by electroplating a thin layer over a high corrosion resistance substrate, such as titanium, niobium, or tantalum. Titanium, niobium, and tantalum form stable oxide layers under anodic condition. Among

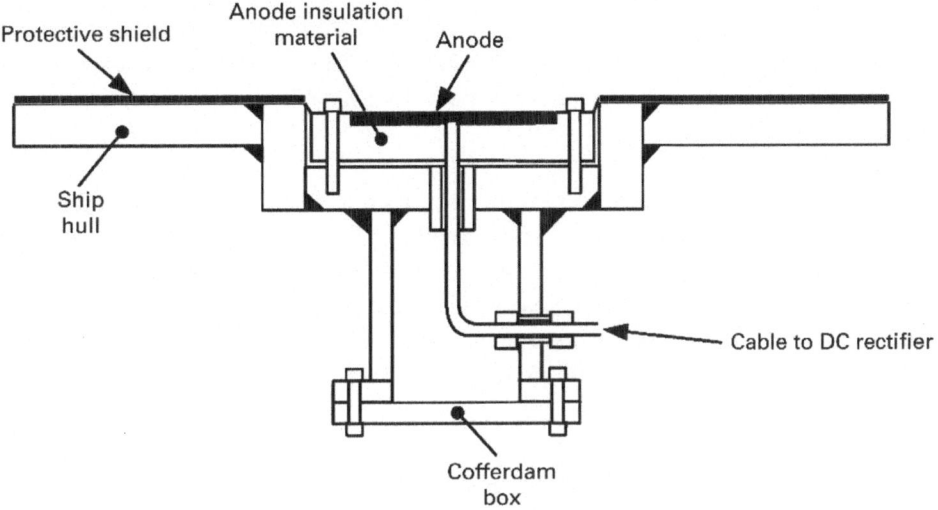

Fig. 4.2 Typical recessed impressed current anode with cofferdam

them, titanium is less expensive. However, it has a much lower breakdown potential than niobium or tantalum. The operating voltage of the platinised titanium anodes is limited by the anodic breakdown potential of titanium due to the dissolution of titanium at un-platinised locations. The anodic breakdown potential of titanium is in the range of 9–9.5 V in the presence of chlorides. Therefore, the maximum recommended operating voltage of platinised titanium anodes is 8 V. The corresponding maximum current density output is approximately 1 kA/m^2 (93 A/ft^2). For cathodic protection systems where, operating volt-ages are relatively high, niobium- and tantalum-based anodes are generally selected. This is because these two substrates have anodic breakdown potentials in the range of 100 V in chloride-containing electrolytes. Platinised anodes are subject to rapid deterioration if the breakdown voltage is exceeded or if the environmental conditions surrounding the anode become acidic.

The consumption rate of platinised anodes is very low, on the order of 6–10 mg/ (A − y) [0.00021–0.00035 oz/(A − y)]. Platinised anodes are available in rod, wire, or mesh form. The rate of platinum consumption has also been found to accelerate in the presence of AC current ripple and current reversal. Most wastage was observed to occur with AC frequencies of less than 50 Hz. The repeated oxidation/reduction processes result in the formation of a brownish layer of platinum oxide. To avoid the occurrence of this phenomenon, a single or a three-phase full-wave rectification is recommended. In addition, the consumption rate of platinised anodes is also adversely affected by fouling, scale, and the presence of certain organic materials such as sugar and diesel fuel.

4.2.2.2 Mixed-Metal Oxide

Mixed-metal oxide (MMO) anodes consist of rare earth oxides baked onto tantalum and niobium or titanium substrate. The consumption rate is on the order of 1 mg/(A–y) [0.000035 oz/(A–y)]. Those anodes are typically available in strip, oval, and circular shaped insulating holders.

- *Liner Loop Anodes.* Produce a powerful output from a relatively small surface area. The anodes are lightweight and easy to install
- *Circular Anodes.* Ideally suited for ships where a smooth hull profile is required. The anodes can be flush mounted in areas where space is available; and
- *Elliptical Anodes.* The elliptical shape enhances current distribution and provides the flexibility to fit into complex hull profiles.

4.2.2.3 Lead

Lead-silver alloys have been used in seawater applications. The lead under anodic current develops a lead dioxide (PbO_2) film that is conductive and prevents deterioration of the lead. The consumption rate of lead-silver alloys is on the order of 91 g/A–y (0.2 lb/A–y). Extruded lead with platinum pins has also been used in seawater applications. The purpose of the platinum pins is to promote the formation of the PbO_2 film. The voltage applied to the anode should not exceed a maximum acceptable value depending on the material of the anode. Recommended maximum acceptable voltages are given in Table 4.1.

The choice of anodes depends on the expected severity of operating conditions together with cost and durability. The current output is inversely proportional to the grounding resistance for a given voltage. This is proportional to the conductivity of the water and can vary by a factor of 100. The anode voltages must be correspondingly raised in poorly conducting waters to achieve the necessary protection current densities. It is not unlikely that the voltage limit of the protection installation and the permissible driving voltages of the particular anodes can be exceeded. In view of this, the protection must be designed for the particular ship and the particular type of water. Short periods in harbour can be ignored. However, in high-resistance waters, under protection can occur because of voltage limitation.

4.2.3 Reference Electrode

A signal from the reference electrode to the controller allows the cathodic protection system to adjust to changing conditions. The controller measures the potential of the hull versus the reference electrode and then signals the power supply to adjust current output for reducing the potential difference between the hull potential and the preset desired potential. The reference electrode is mounted in a domed (i.e., 23 cm (9 in) diameter), circular polyvinyl-chloride holder that electrically isolates the reference cell from the hull.

Table 4.1   Properties of Impressed Current Anodes

Anode materials	Consumption rate g/ A–Y (oz/A–Y)	Maximum current density A/m^2 (A/ft^2)	Maximum voltage (V)
Platinised titanium	0.004–0.012 (0.00014–0.00042)[1]	500–3000 (46–279)	8
Platinised niobium	0.004–0.012 (0.00014–0.00042)[1]	500–3000 (46–279)	50
Platinised tantalum	0.004–0.012 (0.00014–0.00042)[1]	500–3000 (46–279)	100
High silicon cast iron (HSCI)	250–1000 (8.82–35.27)	30–50 (2.8–4.6)	–
Mixed metal oxide (MMO) on titanium substrate	0.0005–0.001 (0.000018–0.000035)	400–1000 (37–93)[3]	8[2]
Lead silver[4]	25–100 (0.88–3.53)	250–300 (23–28)	24

Notes (1) The life of the platinum film may be affected by the electrolyte resistivity, the consumption rate increasing with resistivity. The life of the platinum film can also be affected by the magnitude and frequency of the ripple present in the DC supply. A magnitude lower than 100 mV (RMS) and a frequency not lower than 100 Hz are recommended. (2) In seawater, the oxide film on titanium may break down if the anode potential exceeds 8 V with respect to the Ag/AgCl/seawater electrode. Higher voltages may be used with fully platinised or MMO coated anodes or in less saline environments. (3) In cold seawater, the maximum anode current density of mixed-metal oxide on titanium substrate anodes should be limited to 100 A/m^2 (9.3 A/ft^2) between 0 °C (32 °F) and 5 °C (41 °F) and 300 A/m^2 (27.9 A/ft^2) between 5 °C (41 °F) and 10 °C (50 °F). (4) PbO$_2$ film needs to be formed and maintained by a sufficient anodic current density (typically 100 A/m^2 (9.3 A/ft^2) and 40 A/m^2 (3.7 A/ft^2), respectively, in aerated seawater). The use of platinum pins in the lead/silver alloy may reduce these minimum values (typically to 50 A/m^2 (4.6 A/ft^2) and 20 A/m^2 (1.9 A/ft^2), respectively)

The reference cell is secured to a base or sole plate by a pattern of screws. A series of holes in the reference cell permits passage of seawater, allowing the controller to measure the potential of the hull by the reference electrode. The holes in the reference cell must remain open for the cell to function and are never to be covered by any paint. Reference electrodes should be permanently installed at locations determined by calculation or experience so that the potential of the hull is maintained within the set limits. Normally, two reference cells are installed approximately halfway between anodes powered by the same controller. One acts as a primary control, while the other serves as an auxiliary to verify operation of the primary cell. The auxiliary reference cell is important for verifying system operation on both sides of the hull. It can provide primary service if the first cell fails. Reference electrodes are to be replaceable. They can be designed for electrode change-out while afloat.

The reference electrodes are available in zinc and silver/silver chloride materials. For ice-going ships in arctic waters, similar to the anodes, the reference electrodes attached to the hull and their arrangements should be recessed into the hull with consideration of the heavy mechanical wear from ice.

4.2.4   Wiring and Connections

In ICCP systems, all wiring and connections are to be totally isolated from the electrolyte. Approved cathodic protection dielectrically insulated cable can also be used. Unlike a galvanic anode system, where exposed wire and connections are protected by the anode, any exposed metal in the electrolyte is part of the anode in the ICCP system and will corrode rapidly. The electrical termination between the anode cable and the anode should be watertight, chemically resistant (from nascent chlorine, high pH, seawater, hydrocarbons, and other chemicals), and mechanically robust.

Cables for anodes and measuring electrodes should be laid in pipes or cable banks. The anode cables should be of sufficiently low resistance with the maximum current rating. For various reasons, the permitted voltage drop usually lies between 1 and 2 V. Cables are to comply with appropriate IEC standards for ship wiring and classification society requirements. Cables for impressed current cathodic protection systems are not to pass through tanks intended for low flash-point products. The cable socket to the anodes below the waterline is usually connected via a cofferdam box (refer to Fig. 4.2).

4.2.5   Stuffing Tube(s)

Stuffing tubes for impressed current anodes and reference electrodes are required to pass current to the anodes and control voltage signals from the reference electrode to the controller. If a stuffing tube penetrates a fluid-filled compartment, such as a fuel tank, bilge, ballast tank, or fresh-water tank, the stuffing tube and the electrical cable leading to it must be enclosed in a watertight cofferdam. Requirements for stuffing tubes are addressed in MIL-S-23920, Stuffing Tubes for use with Circular Anodes and Reference Electrodes (Corrosion Preventive). Typically, the stuffing tubes are supplied with the reference electrodes and anodes as assemblies.

4.2.6   Cofferdams

Hull anodes, reference electrodes, and other components that penetrate the hull below the waterline should be designed, constructed, and installed so that the mechanical integrity and watertightness of the ship are ensued. Cofferdams should be used to facilitate the

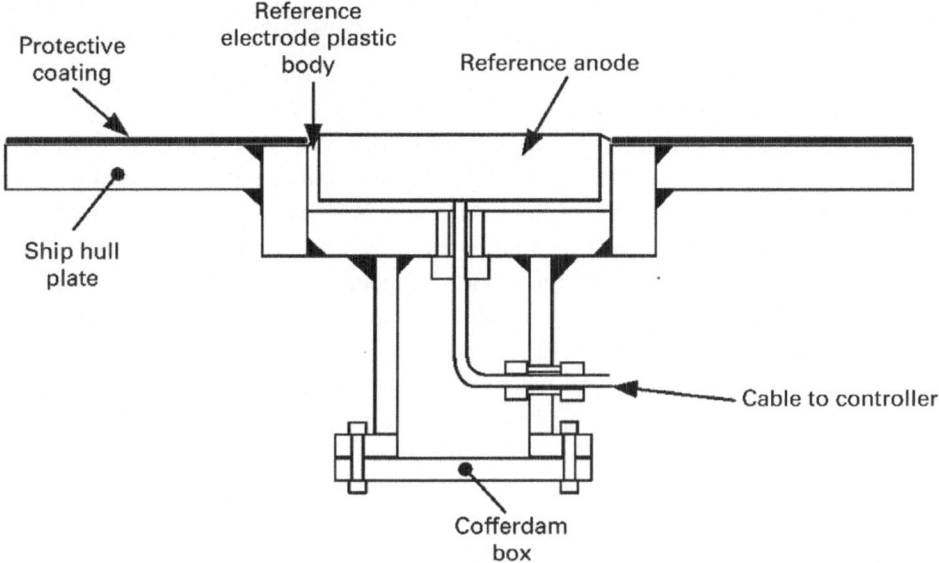

Fig. 4.3 Typical reference electrode mountings with cofferdam arrangement

entry of the cables into the interior of the ship. The construction materials used for cofferdams should be metallurgically compatible to the hull. The steel plate thickness of these cofferdam boxes should correspond to that of the ship hull. The manufacturing and fitting of cofferdams should be in accordance with relevant international, national, or classification society requirements. Figures 4.2 and 4.3 illustrate typical cofferdams of anodes and reference electrode.

4.2.7 Dielectric Shield

The current at which the anodes operate will result in unduly negative potentials immediately adjacent to the anode. These potentials may result in disruption to conventional coatings and can be detrimental to high-strength steels, causing hydrogen embrittlement and a reduction in fatigue life. The dielectric shield is applied to the area of the hull immediately behind and adjacent to the active anode element to prevent shorting of current to the hull and to aid in wider, more even current distribution. The dielectric shield materials selected should be resistant to cathodic disbonding, corrosive chemicals produced at the anodes, as well as significant deterioration or aging. High-performance dielectric coatings are often applied directly to the hull surface prepared in accordance with the coating manufacturers' recommendations. Prefabricated shields are formed from glass (or fibre) reinforced polyester/epoxy or thermoset plastics, either as thin sheet materials pre-bonded to a steel doubler plate or as a direct attachment to the hull. Integral anodes and dielectric shields are available for large flat areas to avoid the need for periodic maintenance of the coatings. As the distance from the anode increases and potential decreases, materials such as high build coatings provide a cost-effective secondary shield area.

The integrity of the shield is critical if the cathodic protection system is to distribute protective current to all areas of the hull. It should have sufficient ductility and toughness. Any damage to the shield will result in unacceptably high potentials on the exposed hull. It normally results in hydrogen evolution at the steel hull surface, which causes either progressive disbondment failure of the shield or potential hydrogen induced cracking of the steel. This area is virtually always in need of maintenance in between the five yearly drydocking. Typically, the standard BS 7361 can be used to help determine the shield dimensions. The size of the dielectric shield will be determined by the shape of the anode, cathodic protection electrical potentials, the maximum current output, and the resistivity of the water. The approximate size of the shield can be calculated from the equations below:

(1) *Disc-shaped Electrode.* The radius, in m (in), of a shield for a disc-shaped electrode is given approximately by the formula:

$$r = \frac{\rho I}{2\pi (E_0 - E)} \quad \text{m(in)}$$

where

E_0 = general potential of the hull when protected, in V

E = most negative potential that can be withstood by the hull paint near the edge of the shield, in V

ρ = water resistivity, in Ω-m (Ω-in)

I = current, in A.

(2) *Strip Anode.* The equipotential lines around a strip anode are approximately elliptical and this is the theoretical shape of an anode shield:

$$b = \frac{L\alpha^{\frac{1}{2}}}{\alpha - 1} \quad \text{m(in)}$$

where

b = minor semi-axis of the relevant ellipse, in m (in)

L = length of the anode strip, in m (in)

$\alpha = \exp\left[\frac{2\pi L(E_0 - E)}{\rho I}\right]$

L_0 = major axis of the ellipse, in m (in) = $\frac{L(\alpha+1)}{\alpha-1}$.

There are possible errors due to the simplified physical model used. For example, the anode is represented as a line source of current, of which the density is uniform along its length, and is regarded as being placed at the surface of a semi-infinite conducting

Table 4.2 Typical minimum distance from anode edge to conventional coating

Anode current output, in A	< 20	20 and < 50	50 and < 150	150 and < 300
Distance, m (in)	0.5 (19.7)	10 (39.4)	1.5 (59.1)	2.0 (78.7)

medium representing the water. In the absence of specific studies and for conventional steels and well-maintained coatings, the minimum distance between the anode edge and the conventional hull coating is given in Table 4.2.

To prevent chemical attack on the hull material and paint directly above an anode due to gas generation, the width of the dielectric shield above the anode should be increased. A minimum width of 100 mm (4 in) should be provided. Procedures for installation and maintenance of dielectric shields are covered in NAVSEA Preservation Process Instruction (PPI) 63,301-001H, *Preservation Process Instruction for Impressed Current Cathodic Protection (ICCP) Anode and Dielectric Shield Inspection/Repair/Replace.*

4.2.8 Arrangement of Anodes and Reference Electrodes

Similar to galvanic anodes, the location and arrangement of impressed current anodes is important to guard against mechanical damage. The essential differences are the decreased number of impressed current anodes needed and the difficulties arising from their connections. Figure 4.4 provides examples of the arrangement of anodes and reference electrodes allowed on the ship. The ideal arrangement is made based on the current distribution used in the design with consideration of coating condition and deterioration during the service life (icebreakers and ice-going ships can have large surface areas in the bow region with no coating).

In contrast to galvanic anodes, there is no fixed rule on the spacing of impressed current anodes, because the current output and range can be regulated. The impressed-current cathodic protection system is designed for a specific ship. In general, the following design criteria should be observed:

(1) For large ships with 150 m (492 ft) or more in length, the stern anodes should be at least 15 m (49.2 ft) from the propeller. This distance can be reduced to 5 m (16.4 ft) on small ships. The reference electrodes should be located where the lowest drop in potential is expected (i.e., distant from the anodes). For large ships the reference electrodes should be at least 15–20 m (49.2–65.6 ft) from the anodes and proportionally nearer in small ships

(2) Anodes are not attached to the rudder but are situated between the rudder shaft and the ship's hull. The propeller is protected via a slip ring on the shaft. Refer to this chapter, Sect. 4.2.7 above

(3) The bow thruster, rudder, scoops, and sea chests are equipped with galvanic anodes

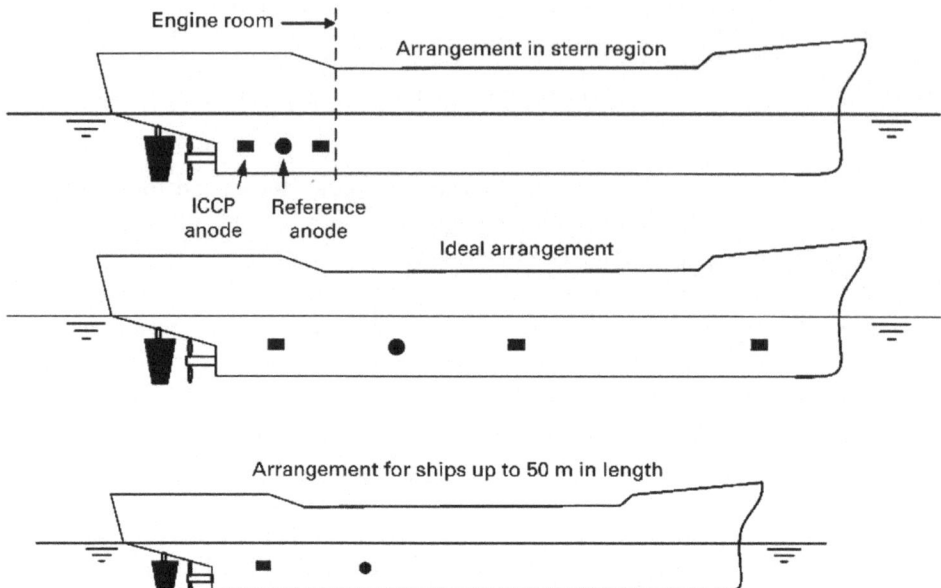

Fig. 4.4 Example of the arrangement of anodes and reference electrodes

(4) The impressed-current system is to be symmetrical (i.e., for the port and starboard sides, the same number of impressed-current anodes and reference electrodes are to be arranged at the same positions). Damage to the ship is to be expected for an asymmetrical arrangement

(5) At least one anode each should be arranged to port and to starboard in the stern area of the ship, preferably in way of the engine room

(6) At least one reference electrode should be arranged for each side. This electrode should be located between the anode and the propeller and as far away as possible from the associated anode (minimum distance approximately 10% of the ship's length)

(7) Ships more than 175 m (574 ft) long should be equipped with a second impressed current system in the bow area

(8) If an impressed current system for the bow area is installed, its control electrode should be located between the anode and the bow; and

(9) A protective shield around the anodes is needed to protect steel and provide a favourable distribution of current. Refer to this chapter, Sect. 4.2.7 above.

The number, dimensions, and location of anodes should be determined in order to be able to deliver the maximum protection current demand I_{max} and to achieve the cathodic protection criteria for the entire cathodic protection zone protected by that cathodic protection system.

4.3 Considerations

The design calculations and specifications should include detailed information on the following:

- Design basis
- General arrangement of the equipment
- Specification of equipment, such as the power source, equipment size, monitoring and control systems, anodes, connection cables, connection and protection devices, reference electrodes
- Specification for installation; and
- Specification for monitoring.

The impressed current cathodic protection system installation should be in accordance with the following:

(1) All electrical connections to the AC or DC electrical system should comply with applicable codes and with the operator's specifications. Nameplate and actual rating of the DC power source should comply with construction specifications. An external disconnect switch in the AC wiring to the rectifier should be provided
(2) Impressed current anodes should be inspected for compliance with specifications for anode material, size, and length of lead wire and to confirm that the anode cap (if specified) is secure. Lead wires should be carefully inspected to detect possible insulation defects. Defects in the lead wires and anode caps are to be properly repaired, or the anode should be rejected. Insulation damage from construction and installation can arise (e.g., due to welding) and should be immediately repaired. Careful supervision of this phase is essential to proper long-term performance of the cathodic protection system
(3) The anode and electrode surfaces should be covered during the coating process (painting, blasting cleaning, and coating application) to protect from coating, dirt, and damage after being mounted during drydocking or construction. Impressed current anodes and reference electrodes should be protected by covers with the visible words "DO NOT PAINT" when installed
(4) Impressed current cathodic protection systems should be installed with negatively grounded or ungrounded electrical systems unless an independent source of power

is provided that is negatively grounded. Connecting the negative to the case prevents inadvertent reversal of polarity

(5) Conductor cable connections to the rectifier from the anode(s) and the structure should be mechanically secure and electrically conductive. Before the power source is energised, verification should be made that the negative (−) conductor is connected to the structure to be protected, that the positive (+) conductor is connected to the anode(s), and that the system is free of short circuits

(6) System controllers should be installed at least 1 m (3 ft) from compasses and other magnetically sensitive equipment. Associated wiring should be either twisted pair or shielded when in proximity to such equipment

(7) A rectifier or other power source should be installed out of the way of operational traffic and remote from areas of extreme heat or likely contamination by mud, dust, water spray, and other contaminants. In areas in which two or more rectifiers are installed, the rectifiers should be spaced for proper flow of cooling air

(8) Openings in the outer shell, such as sea chests, overboard discharges, stabiliser boxes, thrusters, scoops, parts not conductively linked, and shaft penetrations, should be protected additionally with sacrificial anodes

(9) When installing a suspended anode for which separate suspension is required, care should be taken that the lead wire is not in such tension as to damage the anode lead wire or connections; and

(10) Operating personnel should be trained in the function and need for such a system to maintain continued energisation of the system.

Commissioning, Operation, and Maintenance   5

5.1   General

The objectives of the commissioning, operation, and maintenance of the cathodic protection system are:

(1) The cathodic protection system functions continuously in accordance with the requirements of the design and installation; and
(2) The ship's hull, internal surface (ballast tanks), and appendages remain satisfactorily protected from corrosion over the life of the system.

The cathodic protection potential measurements should be conducted by the qualified cathodic protection personnel as stipulated in Chap. 1, Sect. 3.

5.2   Potential Measurement

The protection criteria and effectiveness of cathodic protection systems should be confirmed by direct measurement of the structure potential. Chapter 2, Sect. 3.3 provides the potential measurement techniques and procedure. Ag/AgCl/seawater portable reference electrodes should only be used in undiluted seawater with a salinity of 3.5%. If the potential measurements are undertaken in brackish water, the use of Ag/AgCl/0.5 M KCl electrodes should be considered. However, the potential measurements of galvanic anode systems designed to operate in open seawater should not be done for commissioning purposes in brackish or fresh water. If the vessel is otherwise commissioned in brackish water, the cathodic protection potential measurements for commissioning should be undertaken as soon as possible after the ship is in undiluted seawater.

© The Author(s), under exclusive license to Springer Nature Switzerland AG 2025   63
A. A. Olsen, *Cathodic Protection of Marine Vessels*, Synthesis Lectures on Ocean Systems Engineering, https://doi.org/10.1007/978-3-031-77578-9_5

5.3   Commissioning: Galvanic Systems

For galvanic systems, measurements of the hull/seawater potentials should be undertaken within one (1) month of the installation. The measurement should be done by using a portable reference electrode to supplement any permanently installed reference electrode. The measurement should check that the criteria selected at the design stage in Chap. 2 are met at representative locations. Ballast tank system potential measurements should be made in conjunction with the ballasting programme. A repeat of potential measurements as above should be undertaken one (1) month before the end of the Defect Liability Period (or guarantee) for the ship and its cathodic protection system or within twelve (12) months of drydocking and anode replacement.

5.4   Commissioning: Impressed Current Systems

5.4.1   Visual Inspection

In the drydock, the electrical isolation of the anodes from the hull should be confirmed by electrical resistance measurement. All cable circuits should be checked for continuity and insulation. The polarity of the DC output should be confirmed. The cathodic protection system and all its components should be subject to a complete visual inspection within the drydock to confirm that all components and cables are installed properly, are labelled where appropriate, and protected from any possible damage. The anodes and reference electrodes should be visually inspected to confirm that they are not coated, and the installation is in accordance with the design documentation. The dielectric shield around the anodes should be visually inspected to confirm that the installation is done appropriately as detailed in the design documentation. The shield should be tested for film thickness and for absence of defects ("holidays") both in accordance with the documentation and the dielectric shield manufacturer's recommendations. All inspections data should be recorded.

5.4.2   Pre-energising Measurements

Before the cathodic protection system is switched on when the vessel is floated out from the drydock, pre energising measurements should be made as soon as possible, including:

(1) The potentials of the hull and appendages to seawater with respect to all permanently installed reference electrodes
(2) The potentials of the hull and appendage to seawater with respect to portable reference electrodes; and

(3) Any electronic data logging and/or data transmitting facility installation as part of the performance monitoring system.

5.4.3 Initial Energising

The cathodic protection system is to be energised (switched on) in accordance with the design requirements for initial polarisation. Measurements should be made including:

(1) The potentials of the hull and appendages to seawater with respect to all permanently installed reference electrodes
(2) Confirmation of Polarity. If any steel/seawater potential value shifts in a positive direction, they should be investigated to determine any need for additional testing and/or remedial work; and
(3) The output voltage and current of all transformer rectifiers and the current of all individual anodes.

5.4.4 Performance Assessment

Within one (1) month of energising, potential measurements should be undertaken using a portable reference electrode to supplement the permanent monitoring provisions installed to confirm that the design criteria at the design stage are met at representative locations. A repeat of potential measurements as above should be undertaken one (1) month before the end of the Defect Liability Period (or guarantee) for the ship and its cathodic protection system or within twelve (12) months of drydocking and anode replacement.

5.5   Operation and Maintenance

5.5.1   General

The operation and maintenance testing intervals and procedures should comply with the operation and maintenance manual or as subsequently modified based upon performance of the system for continuous, effective, and efficient operation of cathodic protection systems.

5.5.2   Galvanic Anode Systems

For galvanic anode systems, periodic performance assessment should be undertaken by taking potential measurements at identified locations around the hull. Following the commissioning testing denoted in this chapter, Sect. 5.3, further testing should be performed, typically between nine and twelve months and then at intervals of two to five years, subject to proven design life and planned to drydock intervals. In addition, dependent on the type of ship, location of anodes, and drydocking intervals, a visual inspection of the anodes may be undertaken by underwater inspection. The inspection should assess the consumption of the anodes and check physical damage to anodes. Any damaged, consumed, or missing anodes should lead to potential measurements and estimation of remaining life of anodes for the protection level of the hull or appendages. If necessary, these anodes should be replaced.

5.5.3   Impressed Current Systems

For impressed current systems, normal operation includes confirmation that:

- The system is switched on and all systems are functioning
- A record of the system operation is in place with any down time recorded; and
- All anode current outputs are similar to those settled during the previous assessment.

Overall monitoring and inspection procedures should include:

(1) Measurement and recording of transformer rectifier output total current and voltage (daily)
(2) Measurement and recording of hull steel/seawater potential with respect to fixed measurement electrodes (daily)
(3) Measurement and recording of anode current outputs (daily)
(4) Measurement of parameters from any other sensors installed as part of the performance monitoring system (as appropriate)
(5) Calibration of permanent reference electrodes (annually). A portable reference electrode should be used for their calibration. This should be placed as close as possible to the permanent measurement electrode and the cathodic protection current should be switched off during the calibration procedure
(6) A detailed representative inspection of the entire structure using portable reference electrodes (after drydocking or any major repair/refurbishment of the cathodic protection system and annually); and
(7) The measurement of potential difference between anodes and the hull to verify the metal-to-metal isolation of anode to hull impressed current systems can pose a risk

to divers and are normally switched off during diving operations in their vicinity. If this is impracticable, divers are to be informed so that the necessary actions can be taken with respect to their safety.

5.6 Drydocking Period

Galvanic anodes should be inspected and replaced if their consumption rate is not adequate for the full period to the next drydocking. For impressed current systems, insulating resistance of anodes and electrodes to the hull should be measured. Measurements should be carried out after having cleaned the periphery of anodes and electrodes in order to avoid an electrical continuity due to salt deposits. Insulating resistance should be more than 1 M Ω. Values below 1 MΩ may be acceptable for performance but should be investigated as they indicate possible deterioration. The coating should be visually examined to determine if the coating deterioration is within the value assumed in the cathodic protection design and also if there is any evidence of coating damage caused by the cathodic protection.

5.7 Fitting Out and Lay-Up

5.7.1 General

Precautions should be considered at the design stage to provide adequate protection of underwater surfaces, including appendages, propellers, and shafts of ships, during ship-yard's fitting-out time of newbuilds, refitting time, repair time, or lay-up periods. The choice between suspended galvanic anodes and an impressed current system powered from a shore supply should be determined by accessibility of supply and whether occupation of a berth is anticipated. At permanent moorings, galvanic or impressed current anodes may be laid on the seabed, provided the clearance between the anodes and the keel at low water is sufficient to avoid paint damage.

5.7.2 Fitting-Out Period

The fitting-out period can last many weeks or months depending on the type of ship. Conditions in fitting out berths are often severely corrosive. It is important that cathodic protection is applied during this period to prevent corrosion. If a galvanic anodes cathodic protection system is to be used in service, it should be fitted before launching. If an impressed current cathodic protection (ICCP) system is to be used in service, temporary galvanic anodes should be installed and activated before the ICCP system is installed

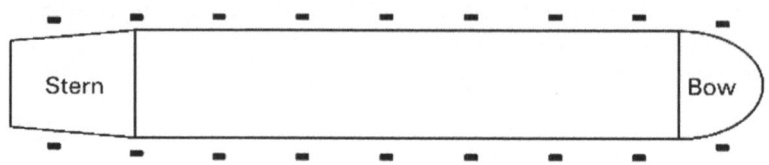

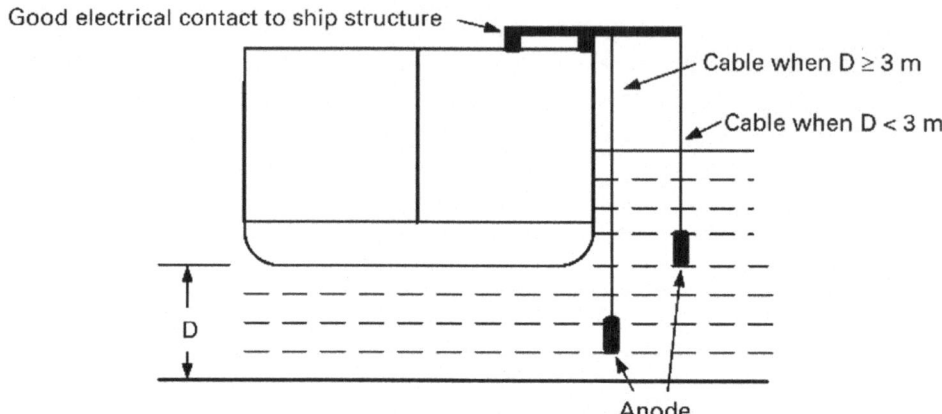

Fig. 5.1 Schematic positions of suspended galvanic anodes for protecting a moored ship

or activated. The temporary galvanic anodes should be bonded electrically to the hull and may be suspended over the ship's side at regular intervals around the ship. Refer to Fig. 5.1. These anodes should be sufficient to provide full polarisation and hull potential should be maintained and monitored at a satisfactory level.

5.7.3 Lay-Up Period

One of the basic criteria in designing the lay-up procedures is the preservation and maintenance of the ship's hull and machinery with appropriate corrosion protection. External coating systems should be in good condition prior to lay-up. The underwater hull area should be adequately protected with suspended sacrificial anodes or the ship's impressed-current cathodic protection system if a continuous power source is available for use. Hull potential should be maintained and monitored at a satisfactory level. At permanent moorings, galvanic or impressed current anodes may be laid on the seabed, provided the clearance between the anodes and the keel at low water is sufficient to avoid paint damage. Ships that are laid up and static for long periods may be subjected to marine growth and microbially influenced corrosion (MIC). In these circumstances, the protection criteria for anaerobic conditions may be used. The initial setting of the correct current requires two or three cathodic protection potential surveys at intervals of a few days. Thereafter, the surveys may be at intervals of several months, provided the water conditions remain stable and the operation of the cathodic protection system is stable. The proximity of other vessels or structures and the need for interaction testing should be considered.

5.8   Stray Current

Underwater hull and shaft corrosion induced by stray electrical currents (undesirable electrical current flows) through water mainly results from improper weld lead hookup and/or from using onshore power. During the fitting-out and lay-up periods, some onboard hot work repairs may be requested. In that case, stray current caused from welding process may cause unpredictable damage to the underwater hull and shaft.

5.9   Interaction Testing

Interaction testing is not normally required of ships due to their mobility. However, if a ship is laid up, or is berthed for long or repeated periods alongside steel quays or jetties, it is recommended that interaction testing should be carried out in order to demonstrate that adjacent structures are not adversely affected by the cathodic protection system. Any changes to the adjacent structure/electrolyte potentials greater than those permitted in EN 50162 should be investigated and corrected. Adjacent structures fitted with cathodic protection are not to have their protection levels changed beyond the levels indicated in EN 50162 by the ship's cathodic protection systems. Adjacent structures not fitted with cathodic protection should not have their corrosion potentials changed by more than + 20 mV by the new cathodic protection system as defined in EN 50162. Similarly, if a ship is laid up or berthed for long periods adjacent to a quay or jetty which is itself protected with cathodic protection, it is recommended that cathodic protection interaction testing be performed to determine the effects on the ship's hull by the adjacent cathodic protection system.

Documentation

6

6.1 General

The design, installation, energising, commissioning, long-term operation, and documentation of the cathodic protection systems should be fully and permanently recorded. This should include all data pertinent to the design, product information, installation, commissioning, and operation and maintenance of the cathodic protection systems. The initial installation documentation should reflect the latest revision of or any change from design specification including the equipment locations, waterline, etc. Commissioning data should include results of the cathodic protection measurements conducted after energising each cathodic protection zone, including:

- Structure potential measurements; and
- Any interaction testing with respect to adjacent structures in the case of laid up vessels.

6.2 Galvanic Anode Systems

The following data should be maintained for reference and should be updated if and when changes are made to the system:

(1) Design criteria including the design life, environment characteristics (i.e., water resistivity, etc.), protection criteria, current density, assumed values of the anodes output current at different periods and working conditions, and the anode's documented ampere-hour capacity and open circuit potential

(2) The number of anodes, their dimensions, weight, specification, alloy composition, documented ampere-hour capacity measured during tests, in accordance with EN 12496 and other characteristics, as well as the manufacturer/supplier references and documentation

(3) The location of each anode as checked during construction, all discrepancies with the design location being highlighted (these locations can be conveniently recorded on a specific drawing of the structure), the method of attachment, and date of installation. This data should be updated during the life of the ship

(4) The location, description, and specification of any current or potential control or monitoring devices, including reference electrodes, measuring equipment, and connecting cables

(5) The commissioning results, including potential measurement data from both fixed reference electrodes (if any) and from a representative inspection of the entire structure using portable reference electrodes

(6) The results of periodic maintenance inspection data including current (if possible) and protection potential measurements, equipment, and the measuring technique to follow the changes of the protection potential status of the ship; and

(7) An Operation and Maintenance Manual which should detail the as-built system, inspection and testing procedures, and inspection and testing intervals. The data detailed above may in addition be incorporated into this volume.

6.3   Impressed Current Systems

The following data should be recorded and maintained for reference and be updated, if and when changes are made to the system:

(1) The design criteria including the design life, environment characteristics (such as water salinity range and resistivity), protection criteria, current density, design values of the anode output current and associated power supply output voltage at maximum current and anticipated operating currents at minimum and maximum extent of coating breakdown

(2) The number and specification of anodes, including their dimensions, anode element composition, connection details, anode current densities and voltages, maximum, average, minimum anode life, as well as the manufacturer/supplier data and documentation

(3) The attachment details of anodes and reference electrodes and the specification of the connecting cables and their through hull arrangements

(4) The location of each anode and reference electrode as confirmed during construction, all discrepancies with design location and the date of installation. This data should be updated during the life of the ship

(5) The specification of any dielectric shield used to include the location, dimensions, surface preparation, material, dry film thickness, and inspection data recorded during the installation of all dielectric shields

(6) The location, detailed specification, drawings, circuit diagrams, and output characteristics of each DC power source (such as transformer rectifiers) with their factory test reports

(7) The location, description, and specification of any performance monitoring and control system, electrical protection devices (fuses, circuit breakers, etc.), measuring equipment, and connecting cables

(8) The commissioning results, including steel/seawater reference electrode potential measurement from fixed reference electrodes, a representative inspection of the entire structure by using portable reference electrodes, current and voltage output values of each DC power source, calibration measurements for each fixed reference electrode and any adjustment made for non-automatic devices

(9) The periodic maintenance inspection results, including steel/seawater reference electrode potential values, DC output values, maintenance data on transformer rectifiers and downtime periods in order to follow the changes of the cathodic protection system status for the structure; and

(10) An Operation and Maintenance Manual which details the as-built system, inspection and testing procedures, inspection, and testing intervals, and provides a fault-finding guide.

Bibliography

ASTM B418-12 (2012) Standard specification for cast and wrought galvanic zinc anodes

BS EN 12474 (2001) Cathodic protection of subsea pipelines

BS EN 16222 (2012) Cathodic protection of ship hulls

EN 12496 (2013) Cathodic protection of ship hulls

MIL-A-18001K (2008) Military specification—anodes, sacrificial zinc alloy

MIL-A-2142A(Ships) (1994) Military specification—anodes, corrosion preventive, magnesium alloy, cast or extruded shapes with cast-in cores

MIL-DTL-24779B(SH) (2009) Military specification—anodes, sacrificial, aluminium alloy